The Red Kite's Year

THE RED KITE'S YEAR

Ian Carter and Dan Powell

PELAGIC PUBLISHING

Published by Pelagic Publishing
PO Box 874
Exeter
EX3 9BR
UK

www.pelagicpublishing.com

The Red Kite's Year

ISBN 978-1-78427-200-5   *Paperback*
ISBN 978-1-78427-201-2   *ePub*
ISBN 978-1-78427-202-9   *PDF*

Text © Ian Carter 2019
Illustrations and artist's notes © Dan Powell 2019

A CIP record for this book is available from the British Library

Cover: Adult birds drifting along Beacon Hill.
       Nice flyby while having a cup of coffee. April 2012.

To the many people who have helped with the restoration
of the Red Kite in Britain and Ireland

CONTENTS

CONTENTS

ADULT R.KITE.
RELAXING - TAKEN NR BEACON HILL
June 2015

FOREWORD

A Red Kite circling overhead is a visible symbol of conservation success in the UK. It's a large bird which is now seen in many parts of Britain and Ireland and is still increasing in numbers and range. If you aren't seeing them already where you live then wait a few years and you probably will. But not long ago they were rare.

When I was a lad, starting birding in Bristol, we had to travel to mid-Wales, up in the hills between Tregaron and Llandovery, to stand any chance of seeing this rare bird. Nowadays, I see Red Kites most days from my home in east Northamptonshire – and they are a joy to watch. Their commonness has not dulled my enjoyment of them, with their orange forked tails and bowed wings. They still catch my eye, and I often stop to watch them; I don't think I'll ever think 'just a kite' when I see one. They will always be special – and that's partly because they were once, and in my memory, so rare, but also because they have always been so wonderful.

The fourth Baron Lilford, who lived a few miles down the Nene Valley from my home, wrote the first major book on the birds of Northamptonshire (published in 1895, a year before his death). He didn't have much to say about Red Kites, as his experience of seeing them in this county was limited to three birds in the hard winter of 1837/38, flying above the lawns of Lilford Hall, but the bird hung on into the 1840s or slightly later.

Lilford tells a story of climbing up to a Red Kite's nest in central Spain, near Aranjuez, in 1865. Kites' nests often have a varied collection of gathered items in them, as did this one. And it was on a scrap of newspaper he found in that nest, up that tree, in Spain, that Lord Lilford first read of the assassination of Abraham Lincoln.

These days, when I drive past Lilford Hall, I almost always see several Red Kites. Local children, in the towns of Kettering, Corby and Oundle, as well as in the countryside between, grow up knowing that the birds above their school playgrounds are Red Kites. I have conversations in the street with people in which Red Kites are mentioned with affection and admiration. Something valuable has been restored to this part of England. Previous generations didn't miss Red Kites – they simply didn't know them. But if someone suggested ripping the Red Kites out of our lives again then there would be uproar, because they now form part of our common memories and sense of belonging.

A once-rare bird is now familiar to many people. And they love it! So it is a great subject for a book. And Ian Carter is a kite enthusiast who has had more to do with the recovery of the Red Kite than most. He is the ideal person to write this book, which I think you will enjoy greatly (I did!).

Mark Avery

ACKNOWLEDGEMENTS

A huge number of people have been involved with the conservation of the Red Kite over many decades, and I hope they will forgive me for not listing them all here. It would be a list stretching to several pages. Initially, conservation efforts centred on the remnant population in mid-Wales, work that developed into one of the world's longest-running species recovery projects. There were many setbacks along the way but gradually the population recovered and, in recent decades, the recovery has gathered pace. Many groups have been involved with the Welsh Red Kites over the years, most notably the Royal Society for the Protection of Birds (RSPB) and the Nature Conservancy Council (now Natural Resources Wales). More recently the Welsh Kite Trust has taken on the mantle, coordinating the monitoring of the expanding population and providing advice on all aspects of this species and its conservation in Wales.

Starting in 1989, efforts have been made to reintroduce the Red Kite into England and Scotland, led initially by the Joint Nature Conservation Committee and then by Natural England (formerly English Nature), RSPB and the Forestry Commission in England, and by Scottish Natural Heritage and the RSPB in Scotland. Of course this work would not have been possible without a source of young birds for release, and the authorities and conservation organisations in Spain, Germany, Sweden and Wales have all helped greatly in identifying suitable donor nests and helping with the collection of nestlings. Following the successes in Britain, work has also been undertaken to reintroduce Red Kites to Ireland, with the Welsh Kite Trust sourcing the birds and the RSPB (in Northern Ireland) and the Golden Eagle Trust (in the Republic) leading on the releases and monitoring of the birds. Many other organisations have been involved in various aspects of the funding and running of the reintroduction programme, including British Airways, Gateshead Council, the National Trust, the Royal Air Force, Yorkshire Water and the Zoological Society of London.

Various voluntary raptor study groups are now involved in monitoring the expanding reintroduced populations, including Friends of the Red Kite (covering the English 'Northern Kites' project), the Southern England Kite Group, Yorkshire Red Kites, and members of the Scottish Raptor Study Groups. These groups make an invaluable contribution to our understanding of the Red Kite and other birds of prey in the areas they cover. Increasingly, as populations continue to expand, it will become more difficult to locate the majority

of the nesting pairs or accurately estimate the size of the local population. Information on rates of population change will then be obtained from national projects such as the British Trust for Ornithology's Breeding Bird Survey and the recently completed national bird atlas.

Countless landowners, farmers and gamekeepers have played a major role in the recovery of the Red Kite through providing secure sites for the release pens, reporting sightings of birds and giving protection to sites used for nesting and communal roosting.

On a more personal note I would like to thank my employers at Natural England (in its former guise as English Nature) for the chance to get involved with the Red Kite in the first place. Having recently left the organisation after almost 25 years as an ornithologist, I look back on those days, working with kites, as some of the happiest and most fulfilling of my career. I would also like to thank Doug Simpson for providing the account of his work on the Yorkshire Red Kite project, and for providing many insightful and useful comments on a draft of the full text. I am grateful to Mark Avery for writing the foreword, Nigel Massen at Pelagic Publishing for taking on this project, and Hugh Brazier for going through the text with a fine-tooth comb and his expert eye for detail.

Finally, I would like to thank my family for encouraging and sustaining my interest in the natural world. My parents allowed me to roam far and wide in the local countryside in a fashion that would be unthinkable now, just four short decades later. By way of proving that point, my two children, Ali and Ben, have had to put up with my company when exploring the wilder corners of the countryside. I hope they derived just a small fraction of the pleasure that I did from those days out in the fields and woods.

Ian Carter, Blagrove Farm, mid-Devon

I would like to thank my parents and brothers for their support, encouragement and transport, our old art-college mucker Anthony Richards for sharing his Carreg Cennen birds with us in the 1990s, Dave Ball for directing us to one of nature's great spectacles, the late Ian Langford for sowing the seeds, and Nigel Massen for making it real. Thanks also to Rosie for sharing the continuing great adventure with me.

Dan Powell, Hillhead, Hampshire

London Hill - Misty Redkite - AM 20/11
Dan Powell

INTRODUCTION

After being a very rare bird for many decades, the Red Kite is making a comeback in Britain and is beginning the process of recovery in Ireland too, both north and south of the border. Thanks to an ambitious reintroduction programme and the continued recovery of the Welsh population it is now, once again, a common bird, at least in some parts of its former range. Although it is still absent from large parts of our countryside, it is spreading to new areas all the time and is becoming a familiar sight in the local countryside for an ever-growing number of people. Things were very different just a few short decades ago. Following centuries of persecution, the Red Kite had long since

The Buzzard and Red Kite are sometimes confused, but given good views they are very different birds. The Red Kite has a more colourful plumage, longer, more rakish wings and a long, forked tail that is constantly twisted from side to side in flight. It is an altogether more elegant bird than the shorter, broader-winged and more powerfully built Buzzard.

been exterminated from England and Scotland. A handful of pairs clung on in the remote uplands of mid-Wales, shrouded in secrecy and subject to intensive protection measures to ensure that the species was not lost completely.

It is hoped that the continued recovery of the Red Kite will bring its fortunes back full circle, for in times gone by this was probably Britain's most common and familiar bird of prey. In medieval times it could be found scavenging on the grimy, unpaved streets of London, as well as filling the skies above a patchwork countryside of woodland and open fields. Its extensive distribution, large size and often slow, languid flight would have made it a familiar bird to almost everyone, and it had a wide range of local names used in different parts of the country. It was *scoul* in Cornwall, *crouch-tail* in Essex, *fork-tail* in Yorkshire

and *boda wennol* (meaning 'swallow-hawk') in Wales. In Scotland and Ireland it had many Gaelic names including *cromán-cearc*. This translates as 'chicken-hawk', hinting at the fact that it was perceived as a threat and, as a result, was persecuted relentlessly.

The Red Kite is not the only large bird of prey making a comeback. The Buzzard too has made a welcome return, spreading naturally to recover ground lost to persecution in central and eastern parts of Britain from its strongholds in the west. For those not used to seeing large birds of prey in the local countryside, the two species can, initially, cause confusion – they both have a large wing-span and both spend long periods gliding or circling over the countryside. But given a little experience and close-up views, the two are very different and are unlikely to be confused for long. The Red Kite is an altogether more elegant and graceful bird, lacking the Buzzard's stocky, compact shape and more powerful build.

The return of the Red Kite has rightly been celebrated. In the Chilterns, there is now 'Red Kite' beer and, for those seeking a healthier option, 'Red Kite' milk. Various local football teams have been named after the bird, and a quick search online will reveal many businesses that have made use of the name – often with a representation of the bird as part of the company logo. People flock to watch Red Kites at sites where they are fed daily and at visitor centres where live pictures are beamed in from a local nest. It is even possible to follow the minute-by-minute fortunes of a Red Kite family in real time on the internet.

Some problems remain, and the continued illegal killing of kites, as well as other birds of prey, especially prevalent in areas managed for intensive grouse shooting, is a major concern. Kites are not universally popular, and they may be blamed for taking songbirds or domestic poultry. Where they have become common in suburban areas they may also be perceived as a threat to small pets and, on occasion, they have even frightened young children with their bold attempts to grab food. Their large size, flocking behaviour and the fact that they spend long periods on the wing has made them an air-safety hazard at certain airfields. If such problems can be addressed there is every prospect that the Red Kite might, once again, become Britain's most common, familiar and best-loved bird of prey. Such a complete recovery will take time but will be well worth the wait and will be testament to just how much our attitudes towards birds of prey have changed.

This book tells the story of the Red Kite month by month. It follows the bird through the ups and downs of the year, from the rigours of raising young during the warm spring and early summer months, to the large and impressive communal gatherings and the struggle for survival in the depths of winter. Separate sections, interspersed with the monthly chapters, deal with subjects such as the history of the Red Kite in Britain (tracing its fall from grace and the beginnings of its human-assisted recovery), the reintroduction programme, threats that it still faces in our modern countryside, and its world status. We hope that our book will help to increase understanding of this impressive bird at a time when more and more people have the chance to see it for themselves.

THE ARTWORKS

A visit to an RSPB nature reserve in the early 1970s was not as straight-forward as it is these days. Then, it involved an exchange of letters with The Lodge (RSPB's headquarters) confirming the date of the visit, a bill, a postal order and the eventual arrival of a permit – how things have changed. And so, with appropriate paperwork in hand, a day at Gwenffrwd-Dinas in the heart of the Red Kite's realm was arranged. At the time, according to my Peterson field guide and the RSPB *Birds* magazine, the only area in the UK where I would have a chance to see this magical beast was mid-Wales. The guide also stressed that Red Kites were so similar to Buzzards that the only way to count a sighting was to see its forked tail – this was going to be tough. My parents, who were always happy to arrange their summer holiday breaks around the growing birdwatching interests of a teenager, provided the means to exchange Hampshire for Wales for a couple of weeks.

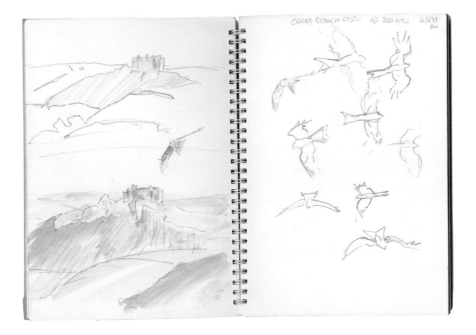

My first sketches of kites that didn't look like pterodactyls.

Gwenffrwd-Dinas was beautiful and kite-free. Devil's Bridge wasn't. Travelling along oak-dressed winding roads we turned a bend and there it was, flying level with the car, so close and big ... Wow! Revealing rich rufous and deep brown wings, contrasting with a white head and bright eyes that stared fiercely at us, making it clear that it owned these skies. The encounter probably only lasted a minute, and the bird soon drifted away, but a lifetime's impression was made – a teenage memory that I've finally attempted to capture on paper (page xiv). And yes, the deeply forked tail was clearly seen.

As far as wildlife was concerned, something of a hiatus soon followed – a gap filled with football, prog-rock, attempts at copying Roger Dean album covers and growing long hair. I reconnected with wildlife at the end of the 1970s. More importantly, though, after studying the artwork of Eric Ennion I connected my own artwork to drawing wildlife – and with it a fledgling career in natural history illustration had begun.

Buzzards, kites and a hare with a sense of humour – how rude!

Ennion's ethos was to observe and draw directly from nature, not to copy slavishly from stuffed birds or from books. From then on many sketchbooks began to fill up with 'odd' bird shapes, though few included those of kites, as opportunities to spend time drawing them were few and far between. The only way to have even a chance of seeing them was to take a trip to Wales – though all that was about to change.

In the late 1990s, with the Red Kite reintroduction scheme well under way, Rosie and I became aware of records in the Meon valley, our local patch. Our time was absorbed exploring the lanes and scanning from the hilltops of this beautiful landscape. To our delight we not only saw kites, we discovered some areas where Brown Hares were allowed to live their lives without being relentlessly hounded by less enlightened landowners. Days when hares and kites graced our views at the same time are special indeed. Then a conversation with a friend steered us in the direction of a winter roost site in the valley, opening up the chance to witness another part of the kite's life, perhaps the most magical of all. We knew of other roosts, bigger roosts, but these were our kites on our patch, making it all the more special to us.

The Red Kite's Year has been some time in the baking. Originally a series of images was commissioned by Ian Langford of Langford Press to be used as chapter headings, the idea being that the rest of the book was to be photographic. However, the more familiar I became with the book, the more it became clear that an illustrative approach would work well, helping to make it more detailed as well as more personal. I also realised that I already had many sketches of the behaviour described by Ian's text – I just needed to find them. All this has resulted in a variety of approaches to the style of artwork appearing in the book. Some are simply scanned straight from the sketchbooks and scraps of paper spanning 40-odd years. Others sit alongside Ian's text, paintings made back in the studio, many also of a personal nature. Sadly, Ian Langford died before this book was published, but I'd like to think he would have enjoyed what we have created.

A bit of a to-do between a Red Kite and a Buzzard – handbags really. At the back of Old Winchester Hill.

Patrolling Red Kites are an increasingly familiar sight in urban areas as well as in the open countryside. They watch over the landscape, constantly on the lookout for feeding opportunities, giving us a chance to watch them.

JANUARY

WINTER FOOD

THE RED KITE IS IN NO SENSE A FUSSY EATER. In fact, it is unrivalled by any other British bird of prey in the wide range of food that it is able to exploit. In the depths of winter, when food may be in short supply, this is a major advantage, allowing the bird to survive on whatever happens to be most readily available in the local countryside. In the late nineteenth century, Lord Lilford gained a vivid impression of the Red Kite's catholic diet through keeping birds in captivity. In his book *Notes on the Birds of Northamptonshire and Neighbourhood* (1895) he describes how:

> *Literally nothing that any bird will eat comes amiss to the Kite ... I have seen a Kite devour rotten cabbage-stalks, scraps of bread, potatoes, fish, flesh, and fowl, fresh, high and putrid, and complacently swallow pieces of stiff leather.*

First and foremost, the Red Kite is a scavenger and is constantly on the lookout for dead birds and mammals that have not been lucky enough to make it through the winter. Small creatures are favoured, as the Red Kite is rather clumsy and awkward on the ground and much prefers to use its aerial agility to snatch up food items that can be carried away to a secure perch and consumed at leisure. Mice and voles may be taken in this way, as are a whole range of familiar farmland birds such as thrushes, finches, buntings and pipits. Rabbits are also a very important source of food and, in some areas, gamebirds such as Pheasants and Red-legged Partridges are eaten frequently.

The Red Kite's superb eyesight comes into its own when picking out small food items, and this is no better demonstrated than in its predilection for

Just the long, bare tail and feet are all that remains visible of this mouse – and they too are just about to disappear. Most small food items are swallowed whole, with the indigestible parts such as fur and bones being coughed up later in pellets.

earthworms. In damp conditions, especially early in the day, earthworms come to the surface – and suitable fields act as a magnet for hungry kites. They can be watched flying back and forth repeatedly, carefully scanning the ground below, and dropping down rapidly on folded wings as soon as food is sighted. Valley sides are often preferred, as the wind deflecting from sloping ground can be exploited to remain in the air with a minimum of effort. Although the Red Kite does not have the ability to hover for long periods (a technique used frequently by Kestrels and, less often, by Buzzards) it is able to hang almost motionless by facing into a strong wind, and this allows it to scan the ground below in detail. Earthworms and other prey are also taken from recently ploughed fields. Farmers have noticed that Red Kites will sometimes follow the tractor as it ploughs a field, joining in with the flocks of gulls and other birds to take advantage of the sudden glut of food that is brought to the surface.

The Red Kite is not a strong and powerful bird and finds it difficult to break through the tough skin of large animals such as deer and sheep. Nevertheless, where more suitable food sources are in short supply, dead livestock can provide a vital lifeline for hungry birds. In the uplands of mid-Wales dead sheep have long provided food for Red Kites, although the meat only becomes available

The Red Kite is wary and hesitant when on the ground and generally prefers to pick up and carry away small food items for consumption at a secure perch in a tree. It will, however, resort to feeding on the ground if a carcass is too heavy to be carried away.

once more powerful scavengers such as Foxes or Ravens have opened up the carcass, or after a certain amount of decomposition has taken place.

Red Kites are often in competition with other scavenging birds when feeding at an animal carcass, including Buzzards and Ravens in some areas, as well as the more familiar Carrion Crows and Magpies. There can be much squabbling as birds try to get a share of the food, with the smaller birds relying on their agility to steal pieces of meat from the larger and more dominant species. Magpies have even been seen apparently working together, one bird hopping up behind a Red Kite and pulling at its feathers, while others nip in to take advantage of the brief distraction.

COPING WITH SEVERE WEATHER

The trend in Britain over recent decades has been towards milder winters, with few prolonged spells of cold or snowy weather. Although this has no doubt been a great benefit to many birds, the Red Kite is reasonably well adapted to cope with all but the harshest of conditions. We have seen the return of periods

Cold, snowy conditions are a mixed blessing for the scavenging Red Kite. Smaller birds and mammals are likely to struggle for survival in cold weather, providing a welcome source of food. But if the snow is heavy and falls regularly it can blanket the landscape, covering food items and making them harder to find.

Red Kites can sometimes be seen at refuse tips, especially in winter, battling for scraps among the myriad gulls and corvids.

In our modern, busy landscapes, road-kills provide a valuable and reliable source of food for the adaptable Red Kite. Quiet country lanes are preferred, but they can also be seen attempting to dodge the traffic on major roads, even on motorways.

of severe weather in several recent winters, with prolonged periods of sub-zero temperatures and snow cover. This makes life more of a struggle, even for the adaptable Red Kite. Frequent heavy snow can be a particular problem, as it quickly covers food items and makes them difficult to detect. It is likely that during these tough conditions Red Kites become more reliant on food supplied by humans, either deliberately at feeding stations and by householders in their gardens, or inadvertently at rubbish tips and through road casualties. The scavenging lifestyle of the Red Kite no doubt helps at times like these, as the increased mortality that cold conditions bring for many animals adds to the potential food supply. Nevertheless, in central and northern Europe, where winters are often very cold and snowy, most Red Kites head southwest to Iberia at the end of the breeding season to escape the worst of the conditions. And in Wales, pairs breeding on the higher ground often move down into the lowlands for the winter.

THE WATCHER WATCHED 27/1/13

27 January 2013

Freezing! The day found us back with the hares – our first visit of the year. To our relief the recent downpours, snow and general harsh weather hadn't driven them away. Of course they were going to be there, they're hares, they're tough – unlike me. The crop was growing fast, but we could just pick them out hunkered down in the same areas of the field as last year. Six were noted, with some dashing about, but no boxing yet.

Overhead two Red Kites kept us company for prolonged periods (sometimes joined by Buzzards), giving superb views and the chance to make some rusty sketches with cold hands. At one stage they shared the airspace with Skylarks – though not at the same altitude. They were clearly seeking food, but surely they were also making the most of the calmer weather to soar in the clear blue skies. Eventually the flask of hot coffee ran out and its warming effect gave way to the bitter cold. Time to repair to somewhere with a log fire.

SKYLARKS - SHARING THE SAME SPACE.
27/1/13 DS

FEBRUARY

THE FIRST SIGNS OF SPRING

THE ACTIVITIES OF THE RED KITE are probably influenced more by the weather conditions in this month than at any other time of year. A cold and snowy February prolongs the winter, and birds that have struggled to find food may be especially vulnerable if severe conditions persist for much of the month. In contrast, mild conditions herald the impending transition from winter to spring – and Red Kites begin to turn their attention towards the all-important breeding season ahead.

The communal winter roosts where Red Kites gather at the end of each day (see *October*) begin to break up in February, particularly if the weather is mild, as pairs start to spend more and more time at their breeding sites. Pairs that are breeding for the first time must try to find a suitable nesting site that is not already occupied by an established pair. At this stage, pairs may be seen prospecting over a number of different areas before finally settling on the chosen site later in the spring.

THE PAIR BOND AND AGE OF FIRST BREEDING

Established Red Kite pairs usually remain together until one of the birds dies, with the bond being maintained, at least loosely, over the course of the winter. Occasionally 'divorces' are recorded, most often following a season when a pair has failed to rear any young. If one member of the pair then has the opportunity to join up with a more successful bird it may do so in order to improve its chances of breeding successfully in the following season. Most Red Kites breed for the first time when they are two years old. Young birds in their first year are

24 February 2012

Since getting married, apart from the vastly improved roads that take us north to Coventry to visit the in-laws over the last 20-odd years, the other joy has been the increase in the numbers of Red Kites seen along the way. On the stretch of road from the north of Hampshire to the M40 it's not unusual to see many individuals gliding low over the verges in search of carrion. In the past, I've tried to keep mental notes of these majestic birds fresh in my mind and then quickly transfer them to paper when we arrive. But I have struggled to relive the moment and the end results were never really satisfactory.

So, during this journey, I had pencil and pad at the ready to make sketches as we saw them – and good fun it was too. We counted about 12 on this occasion. Some views were fleeting, others stunning, with one or two flying very close to the dual carriageway. With eyes on the birds, not the paper, I was surprised to see that four pages had soon filled with their shapes, some pleasing, others indecipherable. It was impossible to add colour at the time (the roads are not that good), so this was added later. Some feel that the whole sketching/painting process is only really true to life if completed in the field. This is OK if practical, but colours always remain fresh in my mind and I like to let them develop a language of their own. By the way, did I mention that Rosie was driving?

unlikely to breed, and most youngsters will continue to gather at communal winter roosts throughout February. First-year birds that do attempt to breed have little chance of success (though a few do manage to rear young) but they no doubt gain valuable experience, which may increase the chances of breeding successfully in the following year.

COURTSHIP BEHAVIOUR AND CALLING

Despite its supreme aerial ability, the Red Kite does not possess the flamboyant courtship display that is well known in some other birds of prey. Activity is mainly restricted to slow circling above the breeding site, often involving both members of the pair. Flights may continue for long periods, and birds can reach a considerable height, becoming scarcely visible to the naked eye, or even disappearing completely into low clouds. There is sometimes a final flourish once they have descended to lower levels, when they can be seen folding back their wings and plunging almost vertically downwards into the nest wood. This is a very welcome sight for fieldworkers monitoring Red Kites, as the point at which a bird enters the canopy during this display is usually close to where the nest will subsequently be built.

Interactions between Red Kites in spring are not always easy to interpret. Are these two birds paired up and indulging in courtship display, or is one bird acting aggressively towards a potential rival?

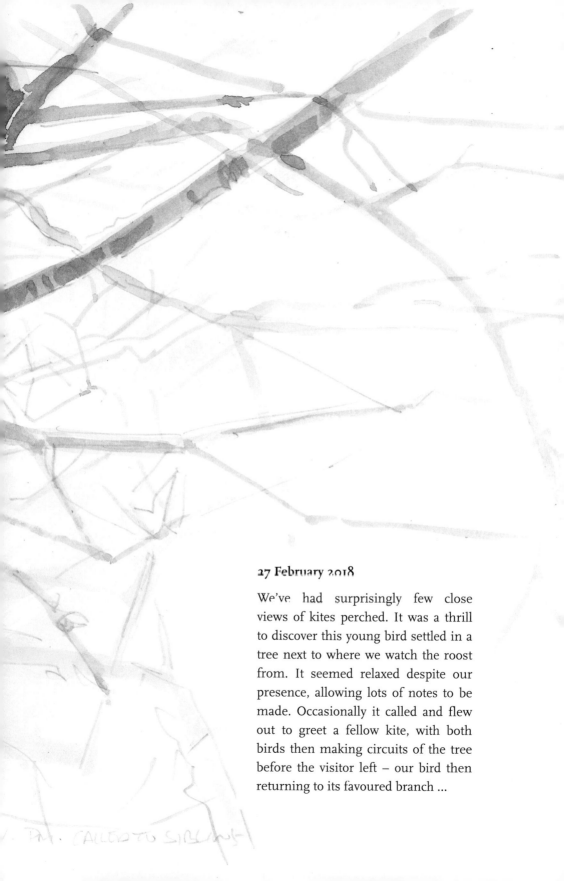

27 February 2018

We've had surprisingly few close views of kites perched. It was a thrill to discover this young bird settled in a tree next to where we watch the roost from. It seemed relaxed despite our presence, allowing lots of notes to be made. Occasionally it called and flew out to greet a fellow kite, with both birds then making circuits of the tree before the visitor left – our bird then returning to its favoured branch ...

The shrill, whinnying calls of the Red Kite are often heard in areas where the bird is common, particularly close to nest sites or where a number of birds are gathered close to a source of food.

Occasionally, Red Kites do indulge in rather more spectacular activity early in the breeding season, including rapid 'rollercoaster' chases, and a highly distinctive flight involving repeated slow, deep wing-beats, often with one bird following closely behind another. It is not always clear whether such interactions involve members of a pair, displaying to help maintain the bond between them, or, perhaps more often, birds involved in a territorial dispute. There is rarely physical contact between rival birds, although this is not completely unknown. Lucky observers have witnessed dramatic mid-air clashes between two Red Kites, with birds thrusting out and then interlocking their talons before spiralling downwards towards the ground. An incident of this nature in Wales ended up with the birds actually crashing down into the woodland canopy below. In the nineteenth century, George Montagu (the man who gave his name to the Montagu's Harrier) described two birds that were even more unfortunate:

> ... so intent in combat, that they both fell to the ground, holding firmly by each other's talons, and actually suffered themselves to be killed by a woodman who was close by, and who demolished them both with his bill-hook.

The Red Kite is not, in general, a noisy bird, and for most of the year it is unusual to hear one calling except where they gather in numbers at a food source or at a communal roost site. In the breeding season, however, they are altogether more vocal, and paired birds often call to each other early in the season when they are together on their breeding territory. The shrill, high-pitched, wavering 'weeee-ooh, ee oo, ee oo, ee oo' call may be made either in flight or from a perch within the nest wood, and is a familiar sound in areas where Red Kites are common breeders. The adults can be especially vociferous when there is disturbance in the nest wood – and human intruders are left in no doubt that they are unwelcome close to the nest. The Red Kite is specially protected in Britain and it is illegal to wilfully disturb birds near to the nest site. Anyone who has inadvertently strayed too close to a nest should heed the warnings from the adults calling overhead and quickly leave the area.

Male or female?

As with many birds of prey (but in contrast to most of the rest of the animal kingdom) the female Red Kite tends to be larger and heavier than the male, although there is much overlap between the two and it is usually impossible to judge the sex of an individual bird. Rarely, when a pair of birds are circling close together, as when displaying over the nest site, it may be apparent that one is a little more lightweight, with a slighter build, than the other. This will almost certainly be the male of the pair.

HISTORY IN BRITAIN

EARLY HISTORY

PERHAPS MORE THAN ANY OTHER BRITISH BIRD, the fluctuating fortunes of the Red Kite have, over the centuries, been inextricably linked with human activities and changing public attitudes. Initially, the Red Kite would have benefited from the increasing impacts of humans on a countryside that had once been dominated by forest. Although woodland is required for nesting and roosting, the Red Kite is primarily a bird of open country, and this is where it finds most of its food. A landscape dominated by dense woodland would have offered few opportunities, but as increasing amounts of forest were cleared to provide land for livestock and crops, so the Red Kite would have prospered. With comparatively little fear of humans, it was also well placed to take advantage of human settlements, and it would have been a regular visitor to farmsteads, villages and even towns, where animal remains and other waste provided a reliable source of food. At one time, the Red Kite was probably our most common and widespread bird of prey and a bird that would have been familiar to almost everyone.

By the fifteenth century, visitors to London were commenting on the high densities of kites and their close association with people, even recounting stories of how they would snatch food from the hands of small children. Such accounts have led some to suggest that these birds must have been Black Kites, which are commonly seen today scavenging in urban areas in parts of Asia and Africa. However, descriptions of the birds from visitors familiar with both species, together with the fact that the London birds were present all year round (the Black Kite is a summer visitor to Europe) makes it clear that they were indeed Red Kites.

Urban areas were attractive to the Red Kite because standards of sanitation in those days were very poor. Animal waste was simply thrown out onto the open streets, providing an abundant source of food. Scavenging by Red Kites, and also Ravens, helped to remove waste and so reduced the risk to human health from outbreaks of disease. In recognition of this street-cleaning role, both these birds were granted special protection in England and Wales under royal statute from the fifteenth century onwards, the first species to be protected for reasons other than hunting. Despite the valued role played by the Red Kite in this period, its scavenging habits also carried darker associations, and it was sometimes seen as a cowardly bird. Shakespeare makes reference to kites many times in his plays, and in *King Lear* the king could think of no better insult than 'Detested kite! thou liest', for the scheming Goneril.

The closely related Black Kite is perhaps the world's most numerous bird of prey. It is a familiar modern-day scavenger in many Asian and African cities – but it was the Red Kite that fulfilled this role in medieval London.

The Raven enjoyed the same legal protection as the Red Kite in the fifteenth century, both species being valued for their role in helping to keep the streets clean by scavenging on refuse and animal carcasses.

A DECLINE IN FORTUNES

The Red Kite's rapid decline from about 1700 occurred for different reasons in urban areas and in the countryside. In towns and cities, the main problem was improving standards of sanitation and a consequent reduction in the amount of food available. By the end of the eighteenth century the bird had gone from the capital, with the last breeding pair being recorded at Gray's Inn in 1777. In rural areas the Red Kite was blamed for taking free-range chickens and even young livestock. Despite its mainly scavenging lifestyle, its large size and the fact that it was often seen feeding at animal carcasses led people to assume that it was a killer and treat it accordingly. Along with other birds of prey and predatory mammals it was increasingly heavily persecuted as rural settlements grew and the human population increased.

The Red Kite was, unfortunately, a very easy bird to kill. Poisoning, in particular, was a highly effective, though totally indiscriminate, means of control. Because the Red Kite is very much a social species, a single animal carcass laced with poison would often result in the death of a number of birds. Traps baited with carcasses or live prey were also sometimes used, and the Red Kite's habit of searching for food by flying slowly, low over the ground, also made it an easy target for the gun. As the quality and reliability of firearms improved so the numbers of birds shot increased. To encourage people to kill birds of prey, bounties of one or two pence were paid by churchwardens for each head presented as evidence. The high number of birds killed demonstrates the immense scale of the persecution, and also gives a clear indication of just how common the Red Kite must once have been. Take, for example, this account by N.F. Ticehurst in the *Hastings and East Sussex Naturalist* (1934), describing the impact of persecution in the Tenterden area of Kent towards the end of the seventeenth century:

> *The Kite ... was evidently a very numerous species and one or more of these grand birds must have been a commonplace everyday sight as they soared over the surrounding forest that formed such a suitable home for them. Between 1654 and 1675 an average of not much more than two a year were paid for, but during the next decade it becomes evident how common the bird must have been, for during this time no fewer than 380 were accounted for, with a maximum of 100 in a single year ... If the same sort of thing was going on in other Wealden parishes, no large raptorial bird could have long withstood such a drain on its numbers and it is no surprise that we no longer have any Kites with us.*

In the nineteenth century, the rapidly dwindling kite population faced renewed attack, this time from a growing army of gamekeepers employed to protect stocks of gamebirds for shooting. Birds of prey and predatory mammals were persecuted with an almost religious fervour. The Marquess of Bute, in 1808, went as far as preparing an oath to be taken by gamekeepers on his estates in Argyll which included the following:

> *... and finally I shall use my best endeavours to destroy all birds of prey etc., with their nests, wherever they can be found therein. So help me God.*

On just one estate in central England no fewer than 183 Red Kites were killed in a ten-year period, along with the following impressive head-count for a wide range of other species.

Predators killed on the Burley Estate in Rutland, 1807–1816

Red Kites	183	Rats	17,108
Buzzards	285	Stoats	1,269
Hawks	340	Weasels	454
Owls	386	Polecats	206
Magpies	1,530	Pine Martens	9
Jays	428	Cats	554
Crows	1,603	Red Squirrels	197
Herons	24		
Woodpeckers	103		

The well-known Victorian passion for collecting led to the death of many birds, particularly birds of prey. The increasing rarity of the Red Kite and its impressive plumage must have made it a particularly attractive target.

As the Red Kite became an increasingly rare species, so the value of its eggs and skins increased, encouraging ever-greater efforts from those wishing to profit from the bird. By the 1840s, the Red Kite was, unsurprisingly, no longer breeding in many English counties and in the following 30 to 40 years it was lost completely from the rest of England and from Scotland. One of the last breeding records in England involved a female recorded as shot from its nest near Bishop's Castle in Shropshire in 1863, while in Scotland, a pair in Caithness in 1884 had its eggs taken and presented to the British Museum. By the end of the 1800s only a handful of pairs survived in remote parts of central Wales, where game shooting, and hence persecution, was less frequent.

THE ROAD TO RECOVERY –
RED KITE PROTECTION IN WALES

Although the Welsh population is thought to have fallen to no fewer that about ten pairs at its lowest point in the 1930s and early 1940s, genetic work has shown that, until recently, modern Welsh birds were all descended from just a single female from this population. So, at the low point, just one pair successfully reared young that themselves went on to breed and make a contribution to the future population – almost as close to extinction as it is possible to get.

From this point onwards, the Red Kite made a slow recovery in Wales, helped by one of the longest-running bird protection schemes that has been carried out anywhere in the world. In the early days one of the main protection measures, overseen by a formal 'Kite Committee', was a system of bounty payments for landowners and farmers with successful breeding pairs on their land – a welcome change from the time described by Ticehurst, when financial rewards had been offered for each Red Kite head. There was a culture of strict secrecy among those involved in Red Kite conservation at this stage, in order to protect the locality of the remaining breeding birds and reduce incidents of deliberate human interference as well as unwitting disturbance by overenthusiastic birdwatchers. Vulnerable nests were sometimes guarded around the clock by volunteers and, at some sites, by members of the armed services.

In the 1980s, although the Welsh population was slowly increasing, levels of breeding success remained very low and there were concerns that this was hindering the prospects for a more rapid increase in numbers. Research showed that only 25–30% of the eggs laid in Wales resulted in young flying from the nest. Many either failed to hatch or resulted in young that died at an early age due to lack of food or the cold, damp conditions often prevalent in mid-Wales during the spring rearing period. After the 1986 breeding season, the Kite Committee decided that a small number of eggs should be taken from the wild, under licence, so that they could be incubated artificially before being replaced in nests as well-grown young. This was a radical step at the time. Indeed, it was the first time in Britain that the eggs of a threatened species had been taken from the wild to be reared artificially and then returned to nests as part of an organised conservation programme. There was opposition from some who saw it as unnatural and unnecessarily intrusive and believed that the birds should be left to recover their numbers naturally.

Provided they were fertile, the eggs taken from the wild stood a higher chance of hatching when incubated artificially, safe from predators and the unwelcome attentions of egg collectors. The eggs were incubated in captivity either in a purpose-built incubator or using a tame bantam hen or Buzzard, and the resulting young were provided with a surplus of food before being returned to nests in the wild. To ensure that the nest from which the eggs were taken remained viable, dummy eggs were used so that the adult birds continued to incubate. Some of these nests were later robbed by egg collectors – who found not the highly prized Red Kite eggs they expected, but replacement fibreglass dummies, or, in some cases, Buzzard eggs.

Between 1987 and 1993, 30 chicks reared in this way successfully fledged from nests in the wild. Many were subsequently found breeding in Wales (identifiable from their wing-tags), clearly demonstrating the benefits of this approach for Red Kite conservation. A longer-term benefit of this work was that it helped to establish the principle of direct human intervention as a useful conservation option for our most threatened and vulnerable species. It is perhaps no coincidence that only two years after the start of the egg manipulation programme in Wales, the first Red Kites were released into England and Scotland at the start of a ground-breaking reintroduction programme.

There have been many setbacks on the long road to recovery in Wales, including the continued depredations of egg collectors and the impact of illegal persecution, especially the use of poison baits aimed primarily at crows and foxes. Poor breeding success as a result of the unproductive landscape and cool, damp climate of mid-Wales has also slowed the rate of population increase. It is all too easy to forget that the kites survived in the Welsh uplands because of their remoteness and the resulting lower levels of human persecution, not because the habitat and weather conditions are ideal. Nevertheless, by the early 1970s, Kite Committee records based on intensive monitoring of the breeding population showed that conservation efforts were finally beginning to pay dividends. By this time the population had struggled to about 25 breeding pairs, increasing to around 30 pairs by 1980 and 60 pairs in 1990. The population increase has been more rapid in recent years, partly as a result of the birds spreading to areas of countryside that are richer in food, making it easier for the adults to rear more young. It is thought that there are now well in excess of 1,000 breeding pairs in Wales, and birds have even managed to spread across the border into western England. The first nest in Shropshire was found in 2006, and since then numbers have increased and the bird has spread into adjacent English counties, something that would have been hard to imagine only 20 or so years ago.

The fortunes of the Red Kite have been further improved in recent decades by the reintroduction programme in England and Scotland (described in a later chapter), and its future in Britain is now more secure than at any stage in the last 200 years.

A slow meander along the Meon valley towards the roost site.

MARCH

FAITHFULNESS TO BREEDING SITES

RED KITES IN BRITAIN TEND TO BE MOST ACTIVE and visible around their nesting area in March, and nest building has often commenced by the middle of the month. This makes it the best time for fieldworkers to locate new breeding pairs. Later in the spring, pairs are more difficult to find. They become more unobtrusive close to the nest site, and even when they are active the increasing cover from leaves on the trees makes them harder to see.

Established pairs tend to remain faithful to a particular nesting area, returning year after year to breed at the same site. Pairs do sometimes move to an alternative breeding site, but movements of more than a few kilometres are rare once a pair has bred for the first time. Breeding territories often remain in use for many years, being taken over by successive generations of birds. The long history of monitoring Red Kites in Wales has shown that some sites, known to be occupied over a hundred years ago, are still in use today.

NEST BUILDING

Red Kites sometimes refurbish the nest that they used the year before, particularly if they were successful in rearing young, and nests used for a number of years in succession can become very large and obvious structures as a result of the new material added each year. Long-established pairs may have several alternative nests from previous breeding attempts and simply select one to renovate for the breeding season ahead. Alternatively, a new nest may be built, either from scratch or using an old Grey Squirrel drey, or perhaps a disused Carrion Crow or Buzzard nest as a secure foundation.

Both adults are involved in building the nest, though it is thought that the male brings in the majority of the material and the female spends more time incorporating it into the nest structure. Substantial sticks of up to 50 centimetres or more are used to build the main nest platform, these either being picked up from the ground or broken from growing trees. The nest may be up to about a metre in diameter and 20–40 centimetres deep.

Nests are usually placed well off the ground, securely lodged in a substantial fork within a mature tree. The majority of nests in Britain are at least 8 metres from the ground, with some as high as 25 metres, representing an arduous climb for licensed fieldworkers wishing to inspect the contents. Inexperienced pairs can sometimes choose rather unpromising situations, and some nests

Nest material being carried to the nest site. Tracking the route taken by birds carrying rags or sticks is a good way to help pinpoint breeding sites.

The adults can seem very clumsy when handling larger sticks, as if they don't have much idea what to do with them. Yet, somehow, they manage to construct a secure nest in which to lay the eggs.

have been found balanced precariously on top of a thin, unstable branch, well away from the trunk. These nests are highly vulnerable in windy weather, and it is not uncommon to find the sorry remains of a nest and its contents beneath the tree after a summer storm. Very occasionally, nests have been built on the ground – these are probably also the work of young and inexperienced birds with little chance of breeding successfully.

Because the Red Kite is a large and long-winged bird, it prefers to nest in fairly open situations where there is a clear aerial route into the site. For this reason, nests are often built in trees close to the edge of a wood or adjacent to a woodland ride or clearing. Small clumps of trees and even isolated trees along a field boundary or in large gardens are sometimes used. Densely planted woodlands, with the trees packed closely together, tend to be avoided, as the birds are reluctant to fly through dense vegetation in order to access the nest.

A typical Red Kite landscape, with patches of woodland providing ideal nesting sites and plenty of open fields for finding food.

MV - 2 KITES OVER THE WOODLAND

... almost always with company ...

NEST SPACING AND BREEDING DENSITY

Unlike many birds of prey, the Red Kite does not defend an exclusive territory from others of its own kind, and potential rivals are only driven away if they stray very close to the nest area. As a result, the nests of adjacent pairs are sometimes built in close proximity. Active nests have been found within 100 metres of each other in Britain, and one pair in the Chilterns had as many as eight other pairs nesting within a 1-kilometre radius. Parts of the Chilterns now support one of the highest densities of breeding Red Kites in the world, with more than one pair per square kilometre across large areas. The highest density of breeding Red Kites ever recorded was in northeastern Germany, where a forest of only 13 square kilometres supported a staggering 136 nesting pairs.

Such high densities have led some people to suggest that the Red Kite nests in loose colonies, much like its close relative the Black Kite. In reality, there does not appear to any advantage to the Red Kite from nesting colonially, and, in most areas, nests tend to be evenly spaced as far as is possible given the availability of suitable woodland. The high density found in northeastern Germany resulted from birds over a wide area taking advantage of a large central block of woodland, surrounded by a much larger area of open farmland rich in food.

APRIL

FINAL TOUCHES TO THE NEST

FOR THE MAJORITY OF PAIRS, nest building continues well into April, when the all-important soft nest lining is added to provide a secure, dry site in which the clutch can be laid. At this time of year, an adult bird seen carrying dry grass or wool into a likely woodland is a sure sign that egg laying is imminent. Where available in the local countryside, wool is almost invariably used by Red Kites to line the nest, and birds will travel some distance to find it. More surprisingly, in parts of Spain where sheep are absent, dried cattle dung is apparently used as nest lining material.

No sooner had the kite landed than a train of Rooks sidled up — to investigate at first, then to irritate!

APRIL 7th 2018

... *collecting wool, then escorted off the premises.*

Adult bird collecting wool and dry grass, copied by youngsters. Sheep totally unconcerned by all the antics.

ADULT BIRD -
MV.
    COLLECTING SCRAPS OF WOOL.
    DIVING/SWOOPING PLUCKING OFF THE FIELD
YOUNG BIRDS WATCHED AND COPIED.

SHEEP UNCONCERNED!   ADULT DRIFTED WEST
                     TOWARDS NEARBY
                     WOODS.

EGGS AND INCUBATION

Most pairs in Britain lay their clutch of eggs sometime during the first three weeks in April, although a few older and experienced pairs will already have laid by the end of March. In contrast, some first-time breeders may not lay until the end of April or even early May. Eggs are laid at intervals of 1–3 days, and a clutch usually consists of 2–4 eggs. Incubation is carried out mainly by the female, but the male helps out at intervals to allow the female time to feed. Although Red Kites have only a single brood each year they may lay a replacement clutch if their eggs are lost early in the season.

As with many birds of prey, incubation usually starts well before the last egg is laid, with the result that the eggs hatch at intervals rather than all at the same time. The last chick to hatch is at a considerable disadvantage and may not survive if food is in short supply. It may even be attacked by its stronger siblings, hastening its demise if it is already weak from a lack of food. This may seem very harsh, but asynchronous hatching has evolved for a good reason. If food is hard to come by and is shared out between several equally matched

Soft material is at a premium for lining the nest and wool is a firm favourite, where it is available. A Red Kite seen carrying wool, dry grass or other similar material into a nest site is a good sign that eggs will be laid in the near future.

chicks in a brood, then there may not be sufficient for any to survive. But if the older chicks grab more than their fair share then they at least are likely to make it, albeit at the expense of their weaker siblings. When food is plentiful, then all the young have a good chance of fledging.

Nest decoration

One of the habits that the Red Kite has become well known for is its tendency to 'decorate' the nest with all manner of material that it finds in the surrounding area. Other stick-nesting birds of prey, including the Buzzard, embellish their nests with fresh leafy twigs just before laying, but the Red Kite, perhaps in keeping with its scavenging nature, uses scraps of rubbish, rags and pieces of plastic. This habit was well known in Shakespeare's time when the Red Kite had an unfortunate reputation for stealing washing left out to dry to incorporate into its nest – 'When the kite builds, look to lesser linen,' warns Autolycus in *The Winter's Tale*. Fieldworkers studying Red Kites in modern times vie with each other to record the strangest items – and in recent years these have included crisp packets, handkerchiefs, plastic gloves, a pair of tights, tennis balls, golf balls, supermarket bags and even an order of service for a funeral at a nearby church. For a reason that is hard to explain, underwear appears with surprising regularity, and children's soft toys are another firm favourite.

The exact purpose of nest decoration has been the subject of some debate. It is possible that some of the soft items are brought in as lining for the nest. Another possible explanation is that the male bird is simply showing off to the female, demonstrating his ability to find a wide range of potential nest-building material in the local countryside. However, recent research on the closely related Black Kite appears to confirm that such material is added as a means of demonstrating to other kites in the area that the nest site is occupied. The researchers found that white plastic items were most likely to be chosen, which is sensible in terms of both visibility and durability. More experienced, higher-status birds used more nest decoration, and this resulted in fewer territorial skirmishes with other Black Kites nesting nearby, thus saving them considerable time and energy. Nest decoration that becomes detached and falls to the woodland floor can provide fieldworkers with a useful clue as to the presence of an active nest in the canopy above, particularly when the nest itself is obscured by leaves.

Recent research on the Black Kite has shown that nest decoration, consisting of such items as rags and plastic bags, is used to advertise the status of the breeding birds and so helps to keep intruders away. It is likely that it serves a similar purpose in the closely related Red Kite.

Occasionally, there can be an unfortunate downside to this habit, as chicks have been known to become entangled in material brought to the nest by the adults. The coloured (often orange) twine used to hold straw bales together is a particular hazard because it is made up of lots of separate threads which can unravel into a deadly mesh. The threads are very strong, and once they are wrapped around the legs of a nestling it has little chance of breaking free. One small chick at a Scottish nest managed to find its way inside a plastic bag and would surely have perished but for a routine monitoring visit. The chick was extricated from its precarious position, and the plastic bag and other rubbish removed.

Another downside to the inclusion of plastic in the nest is that, in extreme cases, it can make the lining impermeable. Two nests in Yorkshire were found deserted, the eggs standing in shallow pools of water. Several others became so waterlogged through lack of drainage that their weight caused them to fall to the ground, where they resembled small heaps of compost. Plastic littering the countryside is a hazard which the forebears of our current Red Kite population would not have had to contend with.

Defence of the Nest

Although they do not establish exclusive feeding territories, a small area around the nest itself is defended from other Red Kites as well as from other raptors and crows. Red Kites are driven away because they are potential rivals, and the male bird, in particular, will be concerned that a rival male could sneak in and mate with his female. Researchers have found that even decoy birds, painted to look like an adult Red Kite and placed close to the nest, are attacked by the resident pair.

Other birds of prey and corvids are a potential problem for a different reason. They represent a very real threat to the eggs or small young, and are often driven away vigorously to discourage a return visit. In areas where Red Kites and Buzzards are nesting in the same wood there may be frequent aerial interactions between the two, and they are relatively evenly matched – the Buzzard is more powerful but the Red Kite has greater manoeuvrability. Often, the two species learn to tolerate each other and both may manage to rear young in nests in the same part of a wood. At other times there is more hostility, and Red Kite young have even been killed by adult Buzzards. In Wales, Red Kites and Ravens have a rather uneasy relationship, and clashes between these two species have also resulted in the death of young and even adult Red Kites.

Very occasionally, Red Kites have been known to take exception to humans close to their nest and one of the adults has launched an unexpected attack. A licensed fieldworker in the Chilterns was actually struck on the back by an especially determined bird while standing below the nest tree. At another Chilterns site, an individual climbing to the nest in order to examine the contents was alarmed to find an adult bird circling below the level of the canopy, coming to within a few metres and necessitating considerable concentration on his part to continue to focus on the job in hand. Although exceptional, even early writers were aware of such behaviour. William Yarrell wrote the following in his *History of British Birds*, published in 1857:

> *The nest and its contents are sometimes vigorously defended: a boy who climbed up to one had a hole pecked through his hat, and one hand severely wounded, before he could drive away the parent bird.*

Corvids, in this case a Carrion Crow, often mob Red Kites, either trying to steal food or because they see them as a threat. In the breeding season the tables are turned, as crows present a very real threat to Red Kite eggs and small chicks. They are vigorously chased away if they venture too close to the nest.

It is believed that it is generally the female that is more aggressive in defending the nest, and this has been confirmed, on occasion, when the birds could be identified from their wing-tags. It is worth making clear that the Red Kite is generally a placid bird and usually limits its protestations to circling over the woodland canopy and calling repeatedly when humans stray too close to the nest. Attacks, or even mildly threatening behaviour, are the exception rather than the rule.

THE RED KITE
REINTRODUCTION
PROGRAMME

Early reintroduction attempts

IT IS A LITTLE-KNOWN FACT THAT THE RED KITE was the subject of a reintroduction project of sorts back in the 1920s and 1930s, involving eggs and young birds being brought to Britain from elsewhere in Europe. One of the first attempts involved the importation of Spanish eggs to Wales by a Mr C.H. Gowland, a dealer in birds' eggs from Liverpool. As far as can be ascertained from the limited records, more than 70 eggs were imported in 1927–28 with a further batch in 1934–35. On each occasion, they were placed in Buzzard nests close to the border with England in the hope that the surrogate parents would hatch and rear them in the wild. There are reports that at least a small number of young Red Kites did fledge from some of these nests but ultimately the project seems to have been unsuccessful, as records of Red Kites from this part of Wales petered out during the 1940s.

In the 1950s, a further project was initiated by Captain H.A. Gilbert. He first brought in Red Kite eggs from nests in Spain, but when this proved unsuccessful he opted instead for taking young birds. Four nestlings were taken in two years, being held captive in pens over the winter before they were released into the wild in Wales in the following summer. It is thought unlikely that any of Gilbert's birds survived for long in the wild, and recent genetic work appears to confirm that there was no lasting infusion of foreign blood into the Welsh population from the efforts of either Gilbert or Gowland.

PLANNING THE RESTORATION TO ENGLAND AND SCOTLAND

Following its extinction in England and Scotland by the end of the nineteenth century, the Red Kite was seen only as a rare visitor, mainly involving young birds wandering from Wales or stray migrants from the continent with no inclination to settle to breed. The small population in Wales was recovering from past persecution but the rate of population increase was slow and threats from egg collecting and illegal poisoning remained considerable. Without human assistance it seemed likely that the Red Kite would remain a rare sight in Britain away from its Welsh heartland for many years to come. This prospect led conservation organisations, in the late 1980s, to begin to think about reintroducing the species. A project was already under way to reintroduce the spectacular White-tailed Eagle to western Scotland, and conservation organisations decided to adopt a similar approach with the Red Kite. After all, it was believed that the lowland countryside of England and Scotland was ideal for kites, and that the only reason they were not present was the severe impact of human persecution in the past.

Having considered a short-list of potentially suitable sites, the Nature Conservancy Council (now Natural England and Scottish Natural Heritage) and the RSPB decided that releases should begin simultaneously at two sites, one on the Black Isle in northern Scotland and the other in the Chiltern Hills on the border of Buckinghamshire and Oxfordshire in southern England. These sites were believed to provide ideal conditions with an abundant food supply and the patchwork of woodland and open fields typical of areas in Europe where Red Kites were still common.

The second key decision was to agree where the young birds for release should come from. Wales was considered as an option but there were concerns that taking a sufficient number of birds to supply two release projects would have an unacceptable impact on the recovering Welsh population. In the end, a small number of Welsh young were donated to the reintroduction programme in the early years but, for the main source of birds, it was necessary to look elsewhere. Captive breeding was considered, but this option would have been expensive and initial trials showed that the Red Kite was a very difficult bird to breed in captivity. Thankfully, conservationists in Sweden and Spain generously agreed that young birds from thriving populations in their countries could be taken for the reintroduction programme. In later years, birds were also taken from healthy populations in Germany.

It is perhaps worth reflecting that the reintroduction proposals were by no means universally popular. Some doubted that the modern landscape was capable of supporting Red Kites, and others had concerns about the extent of human interference that would be involved. They argued that things should be left to nature and the birds should be allowed to come back naturally, in their own time, however long that took. The counter-argument, of course, is that a landscape devoid of kites is an entirely unnatural situation in itself, resulting solely from human interference in the past. Those involved in reintroducing the Red Kite were merely seeking to rectify the unnatural situation created by our own activities in a less enlightened age.

COLLECTING YOUNG AND CARE IN CAPTIVITY

It is fortunate that Red Kite nests are comparatively easy to locate, being large structures, usually sited high up in a prominent tree. Fieldworkers in the donor countries could therefore identify a series of nests from which young birds could be taken each year. The young were collected as nestlings of about 4–6 weeks old, and in order to minimise any adverse impacts on the donor populations, at least one chick was left in each nest visited. The adults continued to bring food to the nest and had a very good chance of rearing at least a single chick. The young removed from nests were held locally at field stations or raptor centres and fed daily until all had been collected and they were ready to be flown to Britain.

One of the big advantages in taking well-grown young from nests in the wild is that the birds spent their first few weeks of life in an entirely natural situation, in the company of other Red Kite chicks on the nest and fed each day by their parents. This greatly reduced the risk that any unavoidable handling during the few weeks spent in captivity would result in the birds imprinting on humans or becoming tame before they were released. Collecting young at 4–6 weeks also avoided some of the major practical difficulties of rearing very small chicks in captivity. Small downy chicks would have to be kept warm and fed regularly on very small pieces of meat, mimicking the roles that the adult birds would fulfil in the wild. When they reach 4–6 weeks, they are already well feathered and so are able to maintain their body temperature even in cool weather conditions.

Climbing to a nest in central Spain in order to collect young Red Kites for release in England. Red Kites can nest at 20 metres or more above the ground, and although they are relatively easy to locate, the task of collecting nestlings for a reintroduction programme is not to be underestimated.

Nestlings about five weeks old, an ideal age for collecting for the reintroduction programme. At this age the young birds are fully feathered and able to regulate their own body temperature. They are also able to handle food for themselves, thus avoiding the need for intrusive hand feeding by humans.

They are also able to tear food into pieces in order to feed themselves, meaning that the hand feeding of small pieces of meat, with the associated risk of imprinting, could be avoided.

The young Red Kites spent about 6–8 weeks in captivity, held in purpose-built wooden release pens at secluded sites where they were unlikely to be disturbed. The pens were divided into several compartments, each with a nest platform and artificial nest, and several large branches to provide perches. Food was placed on the artificial nests daily, using a small feeding hatch above each nest platform so that the birds could not see people as they were being fed. Finding suitable sources of food for the captive birds was not difficult, and much was provided by local farmers, gamekeepers and foresters as a by-product of their routine pest-control programmes. Rabbits, Grey Squirrels, Woodpigeons, Carrion Crows and Magpies formed the bulk of the diet, and fish heads and offal from a local Salmon-processing factory were used in Scotland. The food was chopped up at first, but as the young developed, larger pieces of food, or even whole carcasses, were provided. This ensured that the birds' diet included skin and bone fragments to provide a good balance of nutrients and minerals, as in the wild.

Young Red Kites perched within their release pen, tucked away in a secluded location where disturbance is minimal. These birds are about ten weeks old and are close to being released to take their first flight in the wild.

Once they were able to fly properly, at about eight weeks old, the young Red Kites showed remarkable manoeuvrability within the pens, being able to turn around in mid-air in order to fly several lengths without stopping to land. This helped to strengthen their flight muscles in preparation for their forthcoming return to the wild. Approaching the release date, the birds were examined by veterinary specialists working with the reintroduction programme in order to make sure they were fit and healthy. Blood samples were taken to help assess the condition of the birds, as well as to allow them to be sexed from their DNA.

RELEASE INTO THE WILD

As with all reintroductions, despite all the meticulous background research and planning, there was an element of uncertainty as to how the released birds would fare in the wild, especially in the early years when the release methods were untested. Those responsible for the birds in captivity often admitted to having mixed emotions on the day of release. On the one hand, there was relief that the birds would at last be allowed to fly free in the wild, hopefully the first stage in their becoming an established and familiar part of the local landscape. But there was also an understandable element of concern. How would the birds cope with their first days away from the safety of the release pens? Would they find sufficient food and suitable roosting sites? And would they manage to avoid threats such as illegal persecution and collisions with road traffic?

The approach adopted for releasing Red Kites was designed to ease the transition to independence as much as possible. Food was placed on the roof of the pens or on adjacent feeding platforms on the day of release and this was topped up daily for several weeks. This mimicked the natural situation in which young birds of prey continue to depend on their parents for food for several weeks following their first flight from the nest.

The young kites were released in groups of up to about ten birds when approximately 10–14 weeks old. Releases were carried out by simply removing the front panel from each of the aviary compartments, to allow the birds to fly free. Those present at such releases, often including local farmers and gamekeepers, were given a lasting memory of the first flight into the wild of these impressive birds – and in some cases became lifelong supporters of the reintroduction programme as a result.

The released birds varied considerably in their behaviour. Some took advantage of the food provided close to the pens for only a short period, moving away from the area and gaining complete independence within just a few days. Others remained in the immediate area of the pens for several weeks, returning regularly to feed. After 2–3 weeks, none of the birds appeared to be totally reliant on the food provided, and so the amount given could be gradually reduced.

In order to help keep track of the birds in the wild, before release they were fitted with coloured, plastic wing-tags with an individual number, letter or symbol, as well as a small radio transmitter attached to the two central tail feathers. Wing-tags are unpopular with many people as they are seen as unsightly and, to some extent, act as a symbolic reminder of human interference with what should be a truly wild bird. They are especially unpopular with photographers, who sometimes resort to airbrushing them out of their pictures. While these views are entirely understandable, it was important that released birds were monitored closely so that as much as possible could be learnt about their

Following extinction, the White-tailed Eagle is now doing well in Scotland, with more than a hundred breeding pairs. It was the subject of a pioneering reintroduction project that began in 1975 involving young birds taken from nests in Norway and released in western Scotland. Further release projects have since been undertaken in eastern Scotland and Ireland in order to help restore the species across more of its historic range. It was also once widespread in England, but recent proposals to reintroduce it here have so far not been taken forward, in large part due to vocal opposition by landowners concerned about livestock predation. The Red Kite reintroduction programme has made use of many of the techniques pioneered by work on this species.

As well as all released birds, some wild-reared nestlings have also been tagged. If the colour of both wing-tags can be seen on a Red Kite then its area of origin and age can be determined (see Sources of further information at the end of the book for links to useful websites detailing tag colours). The tag number, letter or symbol must be read to identify the bird as an individual. Understandably, not everyone appreciates seeing artificial strips of plastic on a wild bird, though they are undoubtedly an invaluable aid to monitoring and can be read from almost a kilometre away through a powerful telescope. Thankfully, as the reintroduced populations have expanded, the need to tag birds has diminished, and few birds now carry these unsightly attachments.

movements and behaviour, and the threats that they might face. The tags had a telephone number on the reverse which allowed dead or injured birds to be reported so that they could be recovered quickly for a post-mortem or for treatment. They do not appear to inconvenience the birds in any way and are usually lost after a few years as the nylon attachments become brittle with age and eventually snap, allowing the tags to fall away.

Radio transmitters allowed birds to be followed even more closely during their first year in the wild. Those attached to the two central tail feathers were lost after around 12 months when these feathers were dropped during the annual summer moult. The tag could often be retrieved through radio-tracking so that it could be recycled for use on another bird. More recently longer-lasting transmitters have been attached using a harness arrangement (like a miniature backpack), and these provide information for several years.

Radio-tracking has proved to be an especially effective means of monitoring the birds, allowing individuals to be identified from several kilometres away even if they could not be seen. It also allows dead birds to be located on the ground, and although it is a grim task, the recovery of such birds has proved invaluable in determining the threats that this species still faces. On a more positive note, transmitters with a longer life have allowed remote breeding sites to be tracked down by homing in on the signal from a bird at its nest. Birds that wandered far away from their release area were occasionally searched for using a light aircraft. From the air, with no obstructions, the signal from their radio-tags could be picked up from as far as 30–40 kilometres away, allowing huge areas to be searched in a relatively short period.

Progress so far

Almost three decades have elapsed since the first young Red Kites were released into the Chiltern Hills and on the Black Isle. Much has happened during this time and the Red Kite is, thankfully, now a good deal more secure as a British bird. The first phase of the reintroduction programme was carried out between 1989 and 1994, when 93 young birds were released at each of the two initial sites. Inevitably, some of the released birds were lost, dispersing away from the area never to be seen again or coming to grief as a result of illegal persecution or other misfortunes. But many remained faithful to the release area, or returned following a period away and, overall, survival rates were high. In the Chilterns, for example, a minimum of 76% of released birds survived their first year, and annual survival rates increased to 93% for older age-classes.

The Red Kite is a highly social species, and it was found that as the number of birds in the release area increased, so a higher proportion of those released in subsequent years remained in the area. In the first year, only 11 birds were released, 5 in the Chilterns and 6 on the Black Isle. All 11 moved away from the respective release areas in the first few months and not a single one ended up contributing to the establishment of a new population. Contrast this with the situation in the fifth year of the project when 13 of the 20 birds released in the Chilterns and 18 of the 24 birds released on the Black Isle remained in the area during their first year. A few examples of some of the more notable movements made by released birds are given on page 65.

A major milestone for any reintroduction is the first successful breeding in the wild, as this helps to confirm that the countryside is suitable, not only for the released birds to thrive, but also for them to be able to rear their own young. In the spring of 1991, only two years after the start of the project, two pairs of Red Kites were found settled at breeding sites in the Chilterns, although unfortunately they did not go on to breed successfully. Better things were to come in the following year. In the Chilterns, no fewer than seven pairs were found on breeding territories and four pairs bred successfully, rearing a total of nine young. And on the Black Isle two pairs were located, with the one successful pair rearing a single chick. The birds reared that year were the first young Red Kites to fledge in England and Scotland for well over a century. Another landmark was reached in 1994 when, in both release areas, young raised locally in the wild in 1992 were old enough to make their first breeding attempts and so contribute to the expanding populations through rearing their own young.

The rate at which the new populations became established in the wild exceeded almost everyone's expectations. In part, this reflected the ideal conditions for the birds in lowland England and Scotland, with high levels of food available throughout the year and a climate far more suited to rearing young than the cool, damp uplands of mid-Wales. The breeding biology of the Red Kite also played an important role. Red Kites often make their first breeding attempt when only two years old, a relatively young age for a large bird of prey, and two or three young is a typical brood size.

Although the number of pairs in the wild increased rapidly, especially in the Chilterns, the birds initially showed a great reluctance to spread out and recolonise new areas of countryside. Instead, they remained concentrated in a relatively small area, close to the initial release sites. Each year there was an increase in the number of pairs, but the new pairs tended to nest close to existing pairs, increasing the breeding density in a limited area rather than greatly extending the range. What little range expansion there was came through a very gradual spread outward from the edge of the core breeding area. By the time of a full population survey in 2000, although the Chilterns population had reached an impressive 112 breeding pairs, all but six of these were within 15 kilometres of the release pens. This was not entirely unexpected, as the Red Kite's poor capacity for recolonising areas of suitable habitat away from others of its own kind was, of course, one of the main reasons for the reintroduction project being necessary in the first place.

Red Kites have returned to a landscape that is much changed since they were last present. Human structures increasingly clutter our countryside – but at least they provide handy perches for surveying the scene.

Young birds, taken from a nest in the wild, make their first flight after spending about six weeks in captivity. Coloured plastic wing-tags and a small radio transmitter help fieldworkers to keep track of released birds and monitor their progress. These first flights demonstrate well the supreme aerial ability of the Red Kite. Released birds, having previously flown only a few metres within the confines of the pen, often circle repeatedly over the release sites, some gaining considerable height before finally descending to perch in a tree and take in their new surroundings.

FURTHER RELEASE PROJECTS

In order to speed up the recolonisation process, it was quickly realised that further releases, in new areas, would be needed. As a result, new projects were started in Northamptonshire in central England in 1995 and near Stirling in central Scotland one year later. Initially, both the new projects relied on the continued import of birds from elsewhere in Europe, but, by 1997, it was felt that young birds could be taken from the rapidly increasing Chilterns population for release elsewhere. In fact, since 1997, the Chilterns has supplied no fewer than 237 young birds for release at four different release sites, three in England and one in Scotland. Smaller numbers of young have also been taken from the Black Isle, for release elsewhere in Scotland.

In all, Red Kites have now been released at a total of nine different sites in England and Scotland as part of the reintroduction programme, five in England and four in Scotland (see table on page 66). Projects have also started up to try to restore the Red Kite to Ireland, with releases taking place both in Northern

Examples of notable movements by released Red Kites

- One of the birds released in central Scotland in July 1997 was seen over 500 kilometres away in Gloucestershire in February of the following year, before returning to its release area by May.

- An adventurous bird released in the Chilterns in July 1990 was found dead in January 1991 almost 300 kilometres away, near Rouen in France, after severe storms.

- An unfortunate young bird from the Northamptonshire reintroduction project was found dead under power lines in north Wales. It had been electrocuted, having made the 200-kilometre journey within two weeks of its release in July 1997.

- A bird released as part of the 'Northern Kites' project near Gateshead in summer 2004 was seen with Welsh birds at the Gigrin Farm feeding station in September of the same year, one of many birds from projects in England and Scotland to join up with Red Kites in Wales.

- A bird released in July 1991 in the Chilterns quickly moved away from the release area and was last recorded alive in August of the same year. As it was not subsequently relocated it was assumed to have perished. Then, in March 2012, the bird was found dead at a site in Wales and identified from its BTO ring. It had almost certainly been breeding in Wales undetected for the previous 20 years.

Ireland and in the Republic, in County Wicklow and near Dublin. Young birds for the most recent release projects have all been taken from populations within Britain, both from the native Welsh population and from the expanding reintroduced populations, highlighting just how successful conservation efforts already undertaken have been for this species. While it is still early days for the new projects, the initial signs are already very promising. Small but expanding populations have become established on both sides of the Irish border, and all the populations in England and Scotland continue to increase and expand their ranges, albeit at differing rates.

THE WIDER BENEFITS OF REINTRODUCTION

The Red Kite is a large and spectacular bird with an attractive plumage, often shown off to full effect when floating low over the countryside and, more

Red Kite populations in Britain and Ireland

Release/study area	Estimate of breeding pairs in 2017
Chiltern Hills, southern England	4,000+
Rockingham Forest, Northamptonshire	400+
Harewood Estate, Yorkshire	126
Gateshead ('Northern Kites')	38
Grizedale Forest, Cumbria	5
Black Isle, northern Scotland	70+
Stirlingshire	80+
Dumfries and Galloway	118
Aberdeenshire	26
Republic of Ireland (releases in Counties Wicklow and Dublin)	85
Northern Ireland (releases in County Down)	20
Wales	1,000+

The Chilterns and Rockingham Forest areas no longer support well-defined, discrete populations, although the highest densities of birds are still concentrated at these sites. Birds have now spread to many counties in the southern half of England in varying numbers, and county bird reports or avifaunas are well worth consulting to get the latest picture.

recently, over villages and even towns and cities. Some people have taken to feeding Red Kites and delight in attracting them to scraps of food put out on the garden lawn – an echo of days gone by when the birds were common scavengers around human settlements. As a result of its impressive size, relatively confiding behaviour and attractive plumage, the Red Kite is generally (though not universally) popular in the areas where it has been reintroduced, with considerable demand for information about this new addition to the landscape. Local initiatives have been set up in several areas in England and Scotland to help raise awareness of local wildlife and habitats, using the Red Kite as the central theme. These include Red Kite trails, guided walks, feeding stations and visitor centres with live CCTV pictures from a nearby kite nest. All attract much public interest and all strive to use the Red Kite's appeal to encourage greater understanding of the needs of wildlife more generally and the importance of protecting what remains of local wildlife habitats.

The continued loss of Red Kites from illegal persecution and accidental poisoning (see later for more details) has caused much public concern because of the bird's popularity, and, as a result, attempts have been made to try to tackle these problems. Local people in the release areas are now far more aware of this problem, and in some cases have put considerable pressure on individuals thought still to be using illegal poison baits. There is also now a greater awareness of how seriously the illegal use of poisons is taken by the authorities. One incident in the Chilterns, for example, resulted in a farmer being fined £4,000 for poisoning a Red Kite. It is hoped that tackling these problems will result in the unnecessary loss of fewer Red Kites in the future, and will also help to reduce losses of other birds and mammals affected by the same problems.

Bringing back the Red Kite – a project officer's perspective

Doug Simpson MBE, Yorkshire Red Kites

Red Kites and Peregrine Falcons are at opposite extremes as birds of prey go – yet in my case they are strongly linked. For many years I have been involved in monitoring the progress of Peregrines in the Yorkshire Dales. This provided me with the experience which undoubtedly helped me land the job of Yorkshire Red Kite Project Officer, which I started back in May 1999.

By early June I had organised the construction of two sets of pens in which the young kites were to be housed until ready for release. The arrival of the first batch of ten young birds was confirmation that we were in business. More soon followed, until we had 20 birds, all sourced from the successfully reintroduced Chilterns population. Following their release, the birds' behaviour became increasingly interesting. Most of them established a base at the Harewood Estate release site, though several went exploring further afield. One of these became our first recorded casualty, its decomposed remains being found, through radio-tracking, in a deep narrow valley in East Yorkshire. Other birds which also went eastwards were more fortunate, a few of them settling in the southern Yorkshire Wolds where they established a satellite breeding population. This exciting development was totally unexpected and presented much more of a radio-tracking challenge than had been anticipated so early in proceedings.

As it was not expected that the birds would start breeding until they were two or three years old, it was with considerable disbelief that I located no fewer than three nesting pairs at Harewood in 2000. This wasn't supposed to happen! One of these pairs, the female of which was an older rehabilitated bird we had released in September 1999, raised two young – a major Yorkshire milestone reached so much earlier than expected. We actually had Red Kite chicks in a nest less than 12 months after our first releases. As formal release projects have now ceased, Yorkshire will go down in history as the only reintroduction site in the UK and Ireland to have had successful breeding in its first year.

It was inevitable that we would suffer casualties from one cause or another. A major road bisects the Harewood Estate and has accounted for the deaths of at least two birds. However, poisoning has been the main cause of death, no fewer than 29 birds having so far died through feeding on poisoned baits placed in the open countryside. Surprisingly, several of these incidents occurred very close to the release site, suggesting that whoever was responsible was slow to learn that scavenging birds, fitted with radio transmitters, were likely to be readily detected.

A further 14 birds are known to have died through feeding on rodents which had been killed by rat poisons. There are several different forms of poison available, and when poisoning victims are tested to establish the cause of death they are routinely found to contain traces of at least two different rat poisons. In one case, a bird which was confirmed as having died through feeding on a poisoned bait also had background traces of no fewer than four different rat poisons in its system.

Since releases began in 1999 there have been 15 incidents of Red Kites being shot. Some survived, but others were dead when found. Some of the survivors only came to light when subsequently killed by poisoning, traces of lead shot being found on x-ray examination. Clearly they had been targeted by shooters at some point and suffered non-fatal injuries. However, 2016 saw a sudden upsurge in this type of offence with six victims being found. Three were already dead and two others had such badly damaged wings that they had to be euthanased. The sixth recovered from its injuries and was successfully released. There was a further shooting fatality in 2017 and three more in 2018. Whether poisoned or shot, it is highly likely that the birds which were found represented only a small proportion of the overall numbers killed.

Looking back over the years of the Yorkshire release project, I can identify a number of highlights which made involvement in the process both special and memorable:

- It was very fortunate that I was in a position to apply for the post of Project Officer when plans for the release of Red Kites in Yorkshire were announced. I had taken early retirement from my job in the civil service.

For many years I had undertaken voluntary bird-of-prey monitoring work in my spare time, never giving a thought to the possibility that I might, one day, take up a second career in nature conservation – and be paid for doing it!

- Having got the job, the question then arose as to where it would be based. The original plan, to have the release site on Yorkshire Water's landholdings in the Washburn Valley, was knocked on the head as a result of protests from the grouse-shooting lobby in Nidderdale and beyond. They believed that released Red Kites would send the grouse in the wrong direction, or cause them not to fly at all, on shoot days. After looking at sites all over the county, the project landed at Harewood – just a 20-minute drive from my home.

- I had been aware, from a visit I had made to the Midlands release site, that large pens were required to house the birds pending their release. On enquiring where the Yorkshire pens were coming from, I was told that it was my job to draw up plans and costings, source the materials and then construct them. This was in mid-May, and the young birds would be available from the Chilterns in just a few short weeks. I have never worked so hard in my life as I did over the ensuing weeks. I had already twigged that it was probably the DIY skills which I had mentioned in my CV which had landed me the job, rather than anything to do with birds. The first pens were ready just in time, the stapling of the mesh on the roof being completed while the first consignment of young birds was actually en route from the Chilterns. I envied my counterparts in subsequent releases in northeast England, Northern Ireland and the Republic of Ireland. They had no pens to make – they used mine. I had made them in sections so that they could be readily dismantled and transported – but the last thing I had expected was that they would end up on the other side of the Irish Sea.

- Mention of the northeast England project – officially known as 'Northern Kites' – brings back memories of the many journeys I made between Harewood and Rowlands Gill, near Gateshead. I had been officially deputed to assist staff and volunteers there in setting up their project. The value which they placed on my contribution became clear when I was the only named individual to be thanked at their official opening ceremony. The birds later showed their appreciation too. The first two to be tagged – the first by myself and the second by their project manager under my supervision – made the downhill journey to North Yorkshire where they settled, paired up with Yorkshire birds and raised young.

Apart from the success of the Yorkshire release project, a major source of satisfaction for me has been in witnessing the public reaction to the birds. They have become an increasingly common sight in both rural and urban areas. They are big and recognisable, and sightings of them are regularly reported on our website (yorkshireredkites.net). I can think of no other development or project in this area which has produced such positive and widespread public acclamation, and I take great pride in having been directly involved.

Hazy light, lazy kite.

THE FUTURE

The Red Kite reintroduction programme has had some critics over the years, and has encountered its fair share of problems, not least the loss of birds as a result of the continued use of poisons in our countryside. But, overall, it has been a huge success, and it is now widely regarded as one of the most successful bird reintroductions ever undertaken. The Red Kite has been successfully restored to several parts of England, Scotland and Ireland, and after a long period when the range was expanding only very slowly it is now becoming far more widely established. There are at least a small number of pairs now breeding in most counties in the southern half of England, and it is expected that further consolidation will be achieved in the coming years.

Ultimately, it is hoped that the Red Kite will repeat the feat achieved naturally by the Buzzard and regain its former status as one of our most common, widespread and familiar birds of prey, even if this takes another decade or more to achieve. Based on the densities of birds already present in some areas and the fact that the majority of the countryside provides suitable habitat for this adaptable bird, it is feasible that Britain alone could ultimately support in excess of 50,000 pairs of Red Kites. This may sound unlikely – and it would represent a British population of almost double the current world population. But this figure has already been exceeded by the Buzzard, and there is no reason to think it will not be achieved by the Red Kite, given sufficient time.

MAY

CARE OF SMALL CHICKS

AFTER 31–32 DAYS OF INCUBATION, the eggs hatch and the tiny, down-covered, chicks emerge into the light for the first time. For two weeks or so they are totally dependent on the adult birds, not only for food but also for warmth. The female rarely leaves the nest at this stage and spends most of her time covering the chicks to prevent them from becoming chilled. Disturbance close to the nest is a potentially serious problem while the chicks are still small, especially in cool or damp conditions, as they can quickly succumb if the female is kept away for any length of time. They are also vulnerable to predators such as birds of prey or crows if the female is not close at hand to defend them. A high proportion of the nests that do not go on to produce fledged young fail at this stage, when the chicks are less than two weeks old.

The chicks are fed mostly by the female. She tears small pieces of meat from prey items and offers them delicately to the chicks, bill to bill. Inevitably, the strongest chick usually pushes itself to the front and is fed first, but the female is very patient – and provided there is sufficient food, all the chicks will be well fed at the end of each feeding bout.

STUDIES OF FOOD IN THE BREEDING SEASON

While the female remains on the nest, the male must find food not only for himself and his partner but also for the growing young. Thankfully, this is a time of year when food tends to be abundant as a result of the large numbers of young and inexperienced animals that are available. Food items that provide a substantial meal but are not too heavy to be carried back to the nest are at

An adult Red Kite, most likely the female, keeps a close eye on its small downy chicks, just a few days old. At this stage they are vulnerable to chilling and a range of potential predators. The male bird tends to do the majority of hunting away from the nest while the female remains in close attendance to provide shelter and protection.

a premium in the breeding season. Young birds and mammals such as Woodpigeons, Brown Rats and Rabbits are ideal, whereas larger animals – which are important at other times of year – are less suitable. They can only provide food for young Red Kites if they can be torn into manageable pieces that are not too heavy for the adult birds to lift.

Researchers use several methods for assessing the importance of different prey species in the diet of birds of prey. Food remains are a very useful source of information, especially in the breeding season when they can be found on the nest, as well as around the base of the nest tree. The ground below some nests becomes littered with feathers and bones during the chick-rearing period. Although this provides very useful information, smaller prey will tend to be under-represented. As the young kites grow larger they become more adept at handling food items, with small birds and mammals swallowed whole and even full-grown rats (tails included) sometimes consumed without leaving anything behind by way of clues for fieldworkers.

Red Kites, in common with most birds of prey and owls, regurgitate the undigested parts of their meals as pellets. Those from breeding adults as well as the growing young can be found at nest sites and provide detailed information on what the birds have been feeding on. Hairs from small mammals such as mice and voles are often found in pellets, as well as feathers from a very wide range of birds. Even the tiny hairs, known as chaetae, from earthworms can be detected in pellets, showing that such small items are not ignored completely in the breeding season.

Only direct observations of foraging birds can provide information on *how* food is obtained – whether, for example, it is taken as live prey or scavenged. Such observations suggest that, when feeding young, although the Red Kite still relies mainly on animal carrion, it also takes a certain amount of live prey where this is readily accessible. Nestling birds are a favourite food for some pairs, especially crows and pigeons – taken from their open stick nests if they are spotted by a Red Kite passing overhead. In many ways, the taking of such prey is not very different to scavenging on animal carcasses, as nestlings are immobile and totally defenceless.

Some food items are stolen from other birds, including other birds of prey and members of the crow family. The Red Kite is very agile in flight and has a surprising turn of speed when the need arises, enabling it to steal from a wide range of raptors when their aerial ability is compromised by the weight of food they are carrying. The Red Kite can also be a victim of food piracy, and food carried in the feet is often tucked away beneath the tail – which makes it

Valuable information on what Red Kites are feeding on can be obtained from food remains within regurgitated pellets, found beneath the nest or at favoured roost sites.

inconspicuous to other scavengers in the area (including other Red Kites), as well as reducing drag so that flight is not unduly hindered.

FORAGING RANGE AND HABITAT USE

In contrast to some birds of prey, Red Kites do not defend exclusive feeding territories from others of their own kind, and areas rich in food may be used by birds from several different nest sites with no signs of hostility between them. As the chicks develop and become better able to feed themselves and less susceptible to predators, the female may spend more of her time away foraging. Nevertheless, studies of individually marked birds have shown that the female rarely strays more than about 1 kilometre from the nest, so she is usually close enough to make a rapid return to the nest site should this be required. The male ranges over a much larger area, travelling up to 4–5 kilometres from the nest in his quest for food. In parts of Europe, foraging flights of as much as 20 kilometres have apparently been recorded, but this must be regarded as exceptional and there is no evidence that birds in Britain travel anywhere near this distance in order to locate food.

The Red Kite relies on its superb eyesight to find food, and so forages almost entirely by flying over open ground. Woodlands do not provide useful foraging habitat unless there are sufficient open areas such as clear-fells or natural clearings. Fields with growing crops are used to some extent, although once the crop reaches a certain height they are less useful as food items on the ground become difficult to see. The Red Kite is, in any case, reluctant to land within dense vegetation where it is unable to see its surroundings and would thus be more vulnerable to ground predators. Grassland is a favoured habitat, as it is often less intensively managed than crop fields and food is more likely to be visible from the air throughout the summer.

As in winter, the Red Kite exploits food sources provided by humans during the breeding season, including scraps from rubbish tips and even food put out specifically for it in gardens. A video camera set up at a nest in southern England captured the incongruous image of an adult bird perched on the edge of the nest with a chicken drumstick. Road-kills are often exploited, and food items found by fieldworkers at nest sites are sometimes squashed almost completely flat, having been rolled over by passing vehicles before they were retrieved by an opportunist kite.

*A fruitless, kite-free day was suddenly transformed by a house-high fly-past –
along with a 150-strong escort of Black-headed Gulls from the nesting colony
a couple of kilometres away at Titchfield Haven. They were no doubt trying to
make sure that the kite did not return to threaten their chicks.*

JUNE

THE GROWING BROOD

THE INITIALLY SMALL AND HELPLESS CHICKS develop rapidly in the nest as long as sufficient food is brought in by the adults. For the first two weeks they are covered in grey-white down, but by the start of the third week the first true feathers are beginning to show. These are sometimes referred to as 'pin' feathers because, at first, just the bare feather shafts are all that is visible. A few days later, the tips of the feathers start to emerge from the shafts, resembling miniature paint brushes and making the chicks look somewhat scruffy due to the mix of brown feathers and down. By about four weeks, feathers cover most of the body and at five weeks only small traces of down are left, mainly on the head but perhaps with a few wisps also on the wings and body.

From about three weeks of age the chicks begin to handle food for themselves, attempting to tear pieces of meat from prey left on the nest. However, they are still entirely reliant on the adults delivering food and, for the larger prey items, they continue to receive help in breaking it up into more manageable pieces.

From an early age, the chicks attempt to avoid soiling the nest by backing close to the edge and squirting their liquid droppings out over the side. This procedure can appear rather comical as the ungainly chick struggles across the uneven nest platform without being able to see where it is going. It is also not entirely without risk – as was shown by live pictures from a camera at a nest in Scotland. Having backed to the edge of the nest, one unfortunate chick slipped at the crucial moment and, despite a desperate attempt to cling on, fell from the nest to the ground more than 10 metres below. Thanks to the camera, its plight was witnessed and fieldworkers rushed to the scene. Remarkably, the chick

By three weeks (left), much of the down has been replaced by brown feathers on the back and wings, and by 4–5 weeks (right), traces of down are largely restricted to the head. Young of this age are not brooded by the adult birds and are able to cope with being left exposed to the elements. They can become somewhat bedraggled in wet conditions but, despite appearances, their well-grown feathers provide effective insulation and waterproofing.

was unharmed, and having been replaced on the nest it continued to develop normally with no further mishaps.

Experienced fieldworkers are able to use the amount of 'whitewash' on the ground below a nest to estimate the approximate age of the chicks above. The older the chicks, the more faeces they produce and the further they are able to project them away from the base of the nest tree. A large amount of whitewash over a wide area indicates a brood of chicks that are probably at least four weeks old.

Despite the efforts of the chicks to keep the nest clean, uneaten food often builds up on the nest platform, which can become a decidedly smelly and unpleasant place (at least from a human perspective). Food scraps attract flies and, once their eggs hatch, maggots start to digest any food not consumed by the young Red Kites. It is thought likely that Red Kites have a strong inbuilt resistance to the diseases associated with decomposing meat, as very few of the young succumb to illness, despite the appalling conditions.

PREPARATIONS FOR THE FIRST FLIGHT

By the time they have reached six weeks of age, young Red Kites spend much of the day standing up on the nest platform, indulging in frequent, vigorous bouts of wing flapping in order to strengthen the flight muscles. On occasion, they even manage to lift themselves a few centimetres clear of the nest, which must surely give them confidence for what lies ahead. They become much more adventurous at this stage and no longer restrict themselves to the nest itself if there are convenient branches which they can easily walk onto. These so-called 'branchers' quickly head back to the nest as soon as one of the adults arrives with food. Young of this age can be surprisingly vocal, and the persistent begging calls sometimes alert fieldworkers to nests that were not located earlier in the breeding season.

Cameras have been set up at Red Kite nests in several areas, providing live pictures at a nearby visitor centre or online. As a result, it has become almost routine to follow Red Kites (and many other species) in real time when they are breeding. This helps people to connect with these birds when they are at their most secretive and difficult to observe in the wild. It has also led to some interesting discoveries that would have been almost impossible to make through traditional fieldwork.

In recent years, radio transmitters have been attached to Red Kite nestlings by fieldworkers when they are about five weeks old, allowing detailed information on patterns of movement and behaviour to be gathered. This transmitter has been attached using a harness arrangement (like a miniature rucksack) and will send out a signal at a known frequency for about 2–3 years before the batteries begin to run down. Using a radio receiver, the signal can be detected at a range of 10 kilometres or more from a high vantage point, allowing the fortunes of individual birds to be followed closely. If the worst happens and a radio-tagged bird dies, there is a good chance that the body can be recovered so that the cause of death can be established. As with the live nest cameras, new technology is playing an increasingly important role in helping scientists to understand the natural world.

JULY

LEAVING THE NEST

MOST YOUNG RED KITES MAKE THEIR FIRST FLIGHT away from the nest when they are about seven weeks old. This is, potentially, a very dangerous time – no matter how much wing flapping birds have done on the nest to build up flight muscles, there is only one sure way to find out whether they are capable of making a sustained flight and landing again on a suitable perch. For some youngsters, all does not go according to plan – they quickly lose height and then, when attempting to land in a nearby tree, they are unable to gain a secure footing and start to drop helplessly towards the ground below. If they are lucky, they will manage to grab out at a branch with their feet and flap vigorously to pull themselves up into a standing position. If they are less fortunate, they may end up on the woodland floor, where they are unlikely to be fed by the adults and are extremely vulnerable to ground predators such as Foxes and Badgers. Taking off again from the ground is very difficult for such unaccomplished flyers, and young Red Kites do not possess the climbing abilities of young Tawny Owls, for example, which often end up on the ground but have a remarkable ability to scramble back up to a secure perch.

The usual guidance for people who find a young bird on the ground is to leave well alone as its parents are likely to be in the area and will soon return to feed it. While this is sound advice for most species, including the majority of common garden birds, it is not the best course of action if a young bird of prey is found in this situation. In order to maximise the bird's chances of survival it should be gently picked up and placed on a perch well above the ground, as close as possible to the point where it was found. Once off the ground it will be safe from predators and will at least have a chance of being able to fly back up into the canopy to take food brought in by the adults. Thankfully,

Tawny Owls often share woodland breeding sites with Red Kites. Their young tend to leave the nest at an early age but are superbly well adapted to scrambling back up to safety should they find themselves on, or close to, the ground. Red Kites do not have this ability, and a grounded young bird is unlikely to survive.

young Red Kites tend to play dead when they feel threatened, and this makes them relatively easy to handle (though gloves are a useful precaution). In this situation, it is also worth trying to contact an organisation involved in Red Kite conservation, as they may be able to send a licensed fieldworker to the site to check on the welfare of the young bird in the days ahead.

TOWARDS INDEPENDENCE

For several weeks after their first flight young birds remain totally dependent on the adults, as they are clumsy in flight and too inexperienced to locate good sources of food for themselves. At some nests both male and female parents continue to bring in food, but at other sites, once the young are flying, this role is performed almost exclusively by the male – at these nests the female plays little or no further part in looking after the young, taking a well-earned break after the exertions of the breeding season. Food is usually brought to the nest platform, which the young birds return to regularly, but may also be left on an old nest nearby or transferred to the young on branches high up in the canopy.

During this time the young make a distinctive wheezing, whistling call, presumably to help the parents locate them. This is invaluable to fieldworkers, especially at sites where nests have not been discovered, as it both confirms that breeding has taken place and, with patience, how many young have been reared. All the time, the young birds are gradually becoming more adept at flying and more adventurous in their behaviour. Two to three weeks after first leaving the nest, they will have progressed from risking only short flights from one tree to another, to circling over the nesting area, often rising to a considerable height before returning to the security of the nest wood. The amount of food that the adults bring back to the nest area tends to decline over a period of about 2–4 weeks. This encourages the youngsters to start exploring further afield and finding their own food as they slowly complete the transition to full independence.

RED KITE ADOPTIONS

A few enterprising young birds resort to a rather surprising behaviour in order to prolong the period in which they can take advantage of food provided by adults. These birds take to visiting the nests of other pairs in the area, usually sites at which the offspring are somewhat younger than themselves. The adult birds appear unable to distinguish between their own young and the interlopers, and provide food for both. The wandering young are essentially 'adopted', albeit for only a short period, by the unrelated adults. Nests visited by young birds are normally within a few kilometres of their own nest site, but distances of up to 8 kilometres have been recorded, showing the lengths some birds will go to for a few easy meals.

Some species, especially colonial nesters where pairs breed close together and the risk of provisioning unrelated young is high, have evolved the ability to recognise their own offspring and drive away any youngsters that are not their own. There is evidence that this occurs in the closely related Black Kite, for example. This ability has clearly not evolved in the Red Kite, where nests are usually well spaced and the risk of being disadvantaged by feeding unrelated young is small in comparison to the risk of mistakenly driving away one's own offspring.

MEASURING BREEDING SUCCESS

Fieldworkers monitoring Red Kites measure breeding success by recording the average number of young birds that fledge from active nests in a given area. In England and Scotland, the majority of pairs rear two or three young, with a small number managing only a single chick and the occasional, highly successful, pair rearing an impressive four young. In 2014, a North Yorkshire pair went one better, raising an incredible five young. Taking into account the small number of pairs that fail to rear any young because the eggs do not hatch or the chicks are predated, an average of around 1.5–2 fledglings are produced in most years for each pair. In Wales, where summers are generally cooler and wetter, and food is often harder to come by, the majority of pairs rear just a single chick and only a very small minority manage to rear a brood of three. Because the life expectancy of full-grown Red Kites is very high, with only a small proportion dying each year, even such a low level of breeding success has been sufficient to enable the Welsh population to increase and spread to new areas.

Four well-grown young rather cramped together in the nest with not a lot of space to move around. This is a regular, if rather infrequent, sight in England and Scotland, with two or three to a nest being the more usual scenario. In Wales most nests have just one or two young; three is rare and four is exceptional. In fact, it took Tony Cross of the Welsh Kite Trust a full 25 years of annual monitoring before he first located a brood of four in Wales in 2011. Three of those birds were taken as a welcome contribution to the Irish reintroduction programme, leaving the sole remaining chick with rather more space than it had been accustomed to.

2 July 2018

Shadows and light

Southward bound. The A34 near Abingdon looking more like a parched Bulgarian landscape than a lush English vista at this time of year. Many views of kites, as expected, with the bonus of several groups of young birds, not long departed from the nest, dancing and diving just above the speeding traffic. Sadly, not all the birds had mastered the dance, with several sorry corpses noted along the verges.

8 July 2018

An adult soared in the
extreme temperatures. It was
leaving a trail of dark 'fluff'
behind it. Clearly carrying,
tearing and eating something
gripped in its talons – a
'takeaway' Jackdaw. Behaviour
we'd never seen before –
always something new to
discover.

8/7/2018 AM
31°C
Coombes
Eastmead

2 - AD SOARING
OUT OVER THE
VALLEY BOWL
ONE CLEARLY
CARRYING
SOMETHING
BITS FALLING OFF

- STARTED
TO PLUCK IT
- A JACKDAW

WHITE BEACON IN THE SKY
A34. 2/7/18

The light was strong, and as birds passed overhead they appeared jet black, with the white panels in the wings shining brightly.

THREATS AND PROBLEMS

RED KITES HAVE THE POTENTIAL TO BE VERY LONG-LIVED, and it is a sad fact that many die prematurely at the hands of humans. The oldest Welsh Red Kite so far recorded was ringed 24 years before it died, and there have been several other birds that have approached 20 years of age. In captivity, where a regular food supply is guaranteed and the risks constantly faced by wild birds do not apply, a Red Kite has lived for over 38 years.

As with all wild birds, Red Kites are, at times, affected by natural factors such as disease, starvation and occasionally predation. But the great majority of deaths in Britain result directly from human activities, some of which are targeted specifically at the Red Kite. While these problems have not prevented the species from becoming re-established in parts of Britain, they have certainly slowed population growth and reduced the rate at which the birds can spread out into new areas.

ILLEGAL PERSECUTION

Despite a substantial decrease in persecution levels since the dark days of the nineteenth century when the Red Kite was wiped out from England and Scotland, the threat has not been removed completely. One of the main methods employed by those too idle to use more time-consuming (but legal) forms of pest control is the use of illegal poison baits. This involves an animal carcass such as a Rabbit being laced with poison and then simply left out in the open where it is likely to be found by scavengers. In many cases, these baits are intended for Foxes or Carrion Crows, which are often blamed for killing gamebirds or young livestock. However, the unfortunate Red Kite is such an efficient scavenger that it is often first to arrive on the scene, with predictable results. The fact that Red Kites sometimes forage in loose groups means that several individuals can be accounted for by a single poison bait. This makes the Red Kite rather more vulnerable to illegal poisoning than most other birds of prey, helping to explain how it was wiped out from large areas in the past and the severity of the impact of poisoning in parts of Britain even today.

Some of the poisons in regular use may not have an immediate effect, and birds will fly some distance before succumbing. They are often found below a tree or next to a wall where they landed before the poison took full effect, causing them to fall to the ground. This makes the detection of the bait and the person responsible for placing it well-nigh impossible.

Evidence from post-mortems suggests that thousands of Red Kites have been killed in this way in Britain in recent decades. The problem is at its worst in the uplands, close to areas managed as grouse moors. For example, in northern Scotland almost half of all young kites end up being killed by poison baits, a high proportion on, or immediately adjacent to, moorland areas managed for grouse shooting. Such a high level of persecution has prevented the Black Isle population from increasing at anywhere near the levels seen in the Chilterns despite the same numbers of birds being released over roughly the same period. The dramatic difference between the two populations (over 4,000 pairs now in the Chilterns, but barely more than 70 around the Black Isle) can be attributed almost entirely to higher levels of illegal persecution associated with shooting estates that the birds encounter as soon as they wander away from the release area at the Scottish site.

Birds of prey, along with other predators, are simply not tolerated by some gamekeepers as they aim to produce as many Red Grouse as possible for shooting parties in the autumn. This has, to some extent, tarnished the whole reintroduction programme on the Black Isle. Those involved continue to get away with these activities because, in contrast to lowland release sites, they are operating in remote areas of upland moorland where the chances of getting caught are low.

Illegal persecution is not limited to the use of poison. Shooting is also a problem, and the Red Kite's slow, languid foraging flights, often low over the ground, and its relative lack of fear of people, makes it a comparatively easy target. It is

This hare carcass was laced with the banned pesticide mevinphos, a poison so toxic that a Red Kite died at the spot where it fed. A second dead kite was found close by, and a third, several hundred metres away, was only located because it had a radio-tag. Hundreds of flies that have landed on the bait have also succumbed. The incident was in a lowland area where, thankfully, levels of illegal persecution have declined greatly over recent decades. This is in stark contrast to some upland areas, where such antisocial behaviour continues to cause major problems for birds of prey. Sometimes baits are laced with multiple substances – a bird found recently in North Yorkshire contained no fewer than five different illegal poisons.

difficult to assess how often Red Kites are shot because those responsible are all too aware that they are acting illegally and, if caught, could face a stiff fine or even a prison sentence. They will do all they can to conceal the evidence, and birds shot and killed or seriously wounded are certain to be disposed of quickly. It is only birds that are injured but manage to evade immediate capture that may come to light if they are subsequently found by someone with a more enlightened attitude.

A bird found dead on a railway line in the Chilterns was found to have two different types of gunshot in its body, indicating that it had been shot, and survived, on at least two separate occasions, before finally succumbing when hit by a train – a perilous and fraught life indeed. As with illegal poisoning, this problem appears to be most severe in northern Scotland. Here it has been estimated from post-mortems that at least 8% of birds end their days in this way. When birds are found shot and injured it is sometimes possible for them to be treated in captivity before being released back into the wild, but where the injuries are extensive there is usually no choice but to have the bird humanely destroyed.

Frustrating though it is, it seems that little can be done to prevent continued illegal persecution, especially in remote upland areas, other than hope for a long-overdue change of attitude – a belated realisation that this kind of behaviour is no longer acceptable, even if it does help to produce slightly higher grouse bags for a small minority in the autumn shooting season. For those keen to see the Red Kite freed from the stifling restrictions of persecution, perhaps the most frustrating aspect of all is the fact that the bird poses virtually no threat to gamebirds and is certainly not a predator of livestock. It is either persecuted in ignorance as a perceived threat, simply because of its large size, or is killed incidentally by poison baits primarily intended for other species that pose a greater threat to gamebirds.

The plight of the beleaguered Hen Harrier has received much publicity recently because its numbers are greatly limited by illegal persecution associated with management for driven grouse shooting. It is on the verge of extinction as a breeding bird in England, though scientists estimate that there is sufficient habitat for around 300 pairs. Numbers in other parts of Britain and Ireland are also far lower than they should be. Because the Red Kite has done so well in lowland areas it is easy to forget that it too is killed with alarming regularity in our uplands.

ACCIDENTAL POISONING

In addition to problems with deliberate, wilful persecution, the Red Kite, along with other predators and scavengers, is also at risk from accidental poisoning from highly toxic modern rat poisons and from lead used in ammunition. This so-called 'secondary' poisoning results from birds scavenging on animals that have themselves been killed by rat poison or by lead ammunition.

Modern rat poisons, or rodenticides, are perfectly legal, provided they are used inside buildings or placed within secure bait boxes – but they are as much as 600 times more toxic to some animals than older poisons based on warfarin. The problem for Red Kites stems from the fact that these poisons are slow to act, something that makes them more effective because rats do not learn to associate feeling ill with the poisoned bait and so continue to eat it. They can therefore die out in the open well away from where the bait was consumed. Once again, the Red Kite is especially vulnerable to this threat because it is a highly efficient scavenger and is often the first bird to locate animals that have been killed in this way. It is also not averse to foraging around villages and farm buildings where rodent poisoning campaigns are frequently carried out. The problem is exacerbated by the fact that the Brown Rat is an ideal size for Red Kites, especially in the breeding season when the adults have young to feed. They are large enough to provide a substantial meal but are usually not so big that they cannot be carried back to the nest to feed growing chicks.

Rodenticides work by preventing the blood from clotting, and so result in death through internal bleeding. Fieldworkers in both England and Scotland have been unfortunate enough to find chicks in the nest bleeding from the beak or from growing feathers as a result of the insidious anticoagulant effects of these poisons.

It is difficult to determine the number of Red Kites that have been killed in this way, as many no doubt die slowly and retreat into cover where they are less likely to be found. What is extremely worrying is that around 60–70% of dead Red Kites analysed by laboratories have been found to contain residues from these poisons, and, because they are slow to break down in a bird's body, a potentially lethal dose can build up gradually through repeated feeding on poisoned rodents.

6 April 2012

We dropped in to watch the hares for a while. Mostly they were in loafing mode, pottering about the big field nibbling and grooming, perfect for leisurely painting and shape collecting. The height of the winter barley is beginning to make it a little harder to make out the 'harey lumps' in the field, but it has also accentuated areas of the crop where the hares have chosen to graze. The favoured patches have distinctly shorter vegetation and are a slightly different tone of green.

I think the locals are now used to coming across this daft couple staring intently at fields, as they now pass us by with a cheery wave. Today, I heard a car coming and waved to the occupant as it went by. I then heard it stopping and two things crossed my mind, the first that the driver was pulling up for a chat and the second that he had stopped only because Rosie was sat in the middle of the lane sketching flowers.

It turned out that the gentleman was the estate manager, and a very interesting conversation ensued. He told us some of the background to the area and the farming methods employed. The key factor in there being such good numbers of hares is that they are not shot – and apart from having to get out of the way of the odd tractor every now and then, they are left pretty much undisturbed. As we chatted a Red Kite flew low over the big field. I expected the usual 'get that vermin 'orf my land', but instead he greeted the bird with a smile. Modern farming and wildlife can sometimes make good bedfellows – even on a shooting estate.

Making our way back to Beacon Hill for a coffee, we noticed two Red Kites circling low over the watercress beds at Warnford. With a screech of brakes, we pulled off the road and bundled out of the car to enjoy these superb birds at close quarters. Sadly, an unnatural gap in the primary feathers of one of the birds suggested that not all landowners in the area have the same enlightened regard for wildlife as the gentleman we met earlier in the day.

WARNFORD

LATER WE SAW
A BIRD WITH HOLES
IN ONE WING
— SHOT AT?

AT THE HARE FIELD 8/4/17
9:30

HUNGERFORD
WHILE CHATTING TO
MICHAEL

A KITE
DRIFTED
OVER

BECAME
ALERT

This is a difficult problem to tackle, as rats are unpopular with many people and can cause serious damage to stored foodstuffs if they are not controlled. In many areas, the less toxic rodenticides such as warfarin are still effective and carry a much lower risk of secondary poisoning. Alternative forms of control, such as trapping, may also be effective in some situations. Where the most toxic poisons must be used, it is essential that frequent searches for dead rodents are carried out, and any animals found must be disposed of safely rather than thrown onto an open dump where they are accessible to scavengers. Failure to make regular searches for dead rodents, or using rat poisons contrary to any of the other instructions on the product label, can lead to prosecution and a heavy fine.

Red Kites are also at risk from lead poisoning. It has been known for some time that lead from fishing weights and from spent shot can poison waterbirds. It is ingested as grit, which many of these birds use to help grind up the plant material that dominates their diet. In the past, Mute Swans were regularly encountered showing the distressing effects of lead poisoning. Lead fishing weights have been banned for some time, as has the use of lead in shotgun cartridges at many of the wetland sites that are important for wildfowl. But lead is still commonly used away from wetlands for killing 'pest' species such as

Dead rats are an attractive food source for the Red Kite, but when they have been killed by poison they represent a very real danger. If carried back to the breeding site, the chicks become vulnerable to the same threat – and extreme cases have resulted in the death of all the young in a nest.

The shooting community often talk about having respect for their quarry, and emphasise the links between shooting and conservation. These claims may appear rather hollow when a known poison (with well-documented effects) continues to be pumped needlessly into the environment in huge quantities.

Woodpigeons and Rabbits, and for shooting millions of gamebirds every year. The Red Kite, and other scavengers, are vulnerable when they feed on animals killed in this way and left out in the open countryside.

A recent study of tissue samples from a large number of Red Kites found dead in England found that 37% had high lead levels and 9% had levels high enough to have probably caused the death of the bird. Other predatory species are no doubt also affected, although they have yet to be studied in detail. Given the availability of non-toxic alternatives to lead ammunition, it is surely time to adopt these for all shooting rather than just for shooting over wetland sites, as has already happened in many other parts of the world.

EGG COLLECTING

The eggs of the Red Kite have long been highly sought after by collectors, and this is one of the factors that prevented a more rapid recovery of the Welsh population in the past. In the 1950s and 1960s, when the population was still small and vulnerable, very few years went by without at least one nest being robbed, and in 1985 no fewer than ten nests were robbed, affecting over one-fifth of the breeding population. The reintroduced populations have been relatively little affected by egg collecting, perhaps because collectors prefer to target the native population in Wales and also because, as Red Kite numbers have increased, there has been a corresponding reduction in the 'rarity value' of the eggs. Nevertheless, there is good evidence that at least a small number of clutches have been stolen from nests in southern England.

Following recent changes to the legislation, the penalties for those taking the eggs of protected birds in Britain are now much more substantial – several persistent offenders have even been sent to prison. There are welcome signs that this is beginning to have a deterrent effect, with a reduction in recorded

Changes to the legislation mean that egg collectors now risk a prison sentence if they persist with this outdated and damaging activity. As shown by several recent prosecutions, some people are seemingly unable to help themselves and continue to target our rarest and most vulnerable birds. Once caught, their hard-won collections are destroyed and they may face several weeks behind bars – surely sufficient time to come up with a more constructive pastime.

incidents. Although some eggs will no doubt be taken as Red Kite populations increase and spread into new areas, it is hoped that the impact of this activity will be minimal in future.

COLLISIONS AND ACCIDENTS

Small numbers of Red Kites in Britain are found dead or injured at the edge of a road with injuries confirming that they have been hit by a passing vehicle, presumably when they were trying to feed on road-kill. It is perhaps mainly young and inexperienced birds, or individuals suffering from illness, or from lack of food, that are most at risk, as Red Kites are usually very alert when feeding on the ground and usually have no difficulty in flying up at the first sign of danger.

Other forms of transport can also result in the death of Red Kites. Several have been found dead beside railway lines, and one corpse was even discovered on the front of a train at Edinburgh station, having presumably been carried some distance from where it was killed. Aircraft have also been responsible for the death of Red Kites – perhaps not surprising given that the bird spends so much of its time on the wing, often at a considerable height. Several have been killed by microlights in England and, in Wales, one mid-air collision resulted in the cockpit window of a military jet being smashed, the victim being identified from feathers later found inside the aircraft.

The increasing push for renewable energy sources has led to wind turbines becoming an increasingly common sight in many parts of Britain. Opinion is sharply divided on their aesthetic appeal, but it is becoming increasingly clear that they present at least some risk to birds. Despite their sharp eyesight, birds of prey are unable to follow the motion of the fast-moving rotors when flying close to the turbines, and some collisions are inevitable. Recent studies in Germany, where Red Kites are common breeders, suggest that the species may be especially vulnerable to this threat, no doubt because of its aerial lifestyle. It is also possible that scavenging kites may be attracted to wind farms because of the feeding potential in the form of birds that have already been killed.

Out of 389 birds found dead beneath turbines in Germany, no fewer than 62 were Red Kites. In Britain, an increasing number of casualties have been found under wind turbines in all three countries. Many incidents no doubt go

Birds can be killed by wind turbines when they try to fly through what they perceive to be a safe gap between the blades. This is closed at such speed by the rotors that some unfortunate Red Kites have been sliced clean in two by the impact. The increasing number of wind turbines in Britain means that further casualties of Red Kites, and other species, are inevitable. The number of deaths can hopefully be minimised by careful siting of wind farms away from migration routes and concentrations of vulnerable birds.

unrecorded owing to a lack of regular searching for bodies beneath turbines or because the birds are removed by scavengers before they can be found.

Somewhat surprisingly, given their excellent eyesight and highly accomplished flying skills, Red Kites have, on rare occasions, been recorded colliding with stationary objects, including trees and even the side of a house in Wales. More surprising still was an event witnessed by visitors to the feeding station at Gigrin Farm in mid-Wales. Two birds collided in mid-air, resulting in the death of an adult female. Visitors watched in horror as the bird dropped like a stone just a few metres in front of one of the viewing hides.

Red Kites occasionally become tangled up in refuse, either when foraging on rubbish dumps or, in the case of nestlings, as a result of material brought to the nest as decoration by the adults. One chick at a nest in southern England was found to have plastic wrapped so tightly around its leg that it had begun into the flesh. Luckily it was found by fieldworkers on a routine visit the nest and could be taken into care for treatment before being released back into the wild (see *April* for further examples). Full-grown birds can also become entangled in our carelessly discarded rubbish, as with an individual found suspended by baler twine about 12 metres up in a tree in mid-Wales, and a bird in Yorkshire that was unable to take off because of a tangle of baler twine wrapped around its legs. Both birds were rescued, the grounded individual being caught by throwing a coat over it so that the material could be removed before it was released.

POWER LINES

Large birds like the Red Kite are at risk of electrocution when they attempt to perch on electricity poles, especially on poles with transformer boxes, which have a complicated arrangement of wires in a small area. For a bird to be electro-cuted, it must simultaneously touch either two separate wires or a wire and an earthed metal cross-support. If only a single wire is touched, the electricity does not pass through the body of the bird, and birds habitually perch on live wires with no ill effects. Red Kites are sometimes found dead beneath electricity poles with burn marks clearly visible on the wings, talons or beak, confirming that they were killed by electrocution.

The potential for dangerous poles to cause problems was clearly brought home to one fieldworker in central Scotland, following up a report from a member of the public that a Red Kite had been found dead. The remains of no fewer than 43 dead birds were found beneath three adjacent power poles. These were mainly corvids but also included a Kestrel and a Buzzard in addition to the dead Red Kite.

Where problems are identified, it may be possible to reduce the risk by covering dangerous sections of wire with plastic tubing or by providing artificial perches above the wires. Discussions have already been initiated with electricity companies in some areas to try to limit the number of birds that are killed in this way.

12 August 2012

For the first time in ages the weather was decent and we headed off up the Meon valley to see what had been unfolding in our absence. We pulled in to the entrance to a local cricket pitch, and there to greet us was a sky full of tumbling Red Kites, Buzzards and a Raven – I've never seen the thermals so full in my life! Some birds were following a hay baler, with insects and rodents no doubt on the menu. It still brings instant joy to see kites over a Hampshire downland, and I doubt that feeling will ever fade.

A quick look at the 'big field', to check the state of play there, and at last the crops have been harvested. There was too much of a heat-haze to pick out any hare-bear-ears, so an early morning or evening visit is due soon.

Back to Beacon Hill. To walk out on chalk downland in full flower is a privilege and delight. Clustered Bellflower, Autumn Gentian, Marjoram, Round-headed Rampion, Field Scabious, Eyebright and Harebell all added to a wonderful mosaic of colours. Among the flowers danced Chalkhill Blues and tiny Brown Argus butterflies, but sadly we couldn't locate any Silver-spotted Skippers. I hope that we simply missed them and they have not been lost from this site. Several more kites noted soaring in the distance.

AUGUST

Fending for Themselves

BY THE MIDDLE OF AUGUST, most young Red Kites are fully independent. They have moved away from the nest wood and are beginning to explore the local countryside, all the time learning about potential food sources and improving their chances of long-term survival.

It was once thought that young birds were accompanied by their parents for the first few months of life, even after they had left the nest wood. Studies of radio-tagged birds have shown that this is not the case. Young birds are certainly often seen in the company of adults in the autumn, but this simply reflects the Red Kite's highly social nature, particularly when gathering at a food source, and the birds involved are usually not closely related to each other.

The sight of large groups of young birds together with one or two adults has even led to reports of Red Kite crèches, the assumption being that these adults take on responsibility for looking after the young of a number of different breeding pairs in the local area. Once again, such reports are not based on good evidence, and mixed gatherings of unrelated young and adults are to be expected in late summer at sites rich in food.

Plumage differences

Many people are surprised to learn that young Red Kites are about the same size as the adults by the time they leave the nest. The wings and tail may be very slightly shorter in young birds, but the differences are so small as to be barely noticeable. However, it is relatively easy to distinguish young birds from adults based on their plumage if birds are seen well. Young birds are much paler than adults with a duller brown, more 'washed-out' effect, lacking any obvious reddish tones in either the wings or the tail. Pale tips to the wing coverts also form a whitish line across both the under-wing and upper-wing of young birds. In good light this can be a very obvious feature, and it can be seen from a considerable distance away with binoculars or a telescope.

On perched birds the difference in eye colour between the two age groups is a good feature to look for. Adults have bright, clear yellow eyes which can look almost piercing at close range, set against the pale grey background of the head. Young birds have much duller eyes in their first few months, although these do begin to lighten gradually during the course of their first year. Another feature, best seen when a perched bird is facing towards the observer, is the streaking on the breast. Adult birds have breast feathers with dark streaks which stand out strongly against a pale, reddish ground colour. Young birds have a few black streaks but also many pale-edged feathers, giving them a rather drabber and subdued appearance.

The bird to the left (top and bottom) is an adult, with a more rufous tone to the wings and upper body, and a reddish tail. The juvenile bird on the right has a duller appearance and a noticeable line across the upper-wing, formed by pale tips to the wing coverts.

The juvenile bird (on the right) has a much duller eye and a drabber more 'washed-out' appearance to the plumage than the adult.

All these differences can appear rather subtle at first, but, with experience (and good views), it becomes quite easy to separate the two age groups. Late summer is a good time to practise, when the plumage of young birds, not long out of the nest, is still fresh, and mixed groups of adults and young are often found in areas where the Red Kite is common. Plumage differences that seem hard to discern when studying a series of lone individuals are usually much more obvious when a direct comparison can be made between birds of different ages.

THE ANNUAL MOULT

There is an alternative method for distinguishing between young and older birds in late summer and autumn, based on the regular moult which all birds must undergo to replace worn-out feathers. Adult Red Kites have an annual moult during which most of the wing and tail feathers are replaced in sequence. This starts in spring but is a protracted process, taking up to six months, because birds can only afford to have a small number of feathers re-growing at any one time. Small gaps at the same point in each wing may be the only indication that the moult is under way early in the summer. However, by the end of the breeding season, when the duties of rearing young are at an end, the moult proceeds more rapidly – and birds may then look very tatty with substantial gaps in both the wings and tail. In contrast, recently fledged young have only just grown their first full set of feathers (which will not start to be replaced until the following spring) and usually look immaculate – with not a feather out of place.

A VERY TATTY ADULT BIRD IN HEAVY MOULT!
JOINED BRIEFLY BY A YOUNGSTER IN FINE CONDITION

The wings of some Red Kites become so ragged and tatty due to moult in late summer that it is surprising they can still manage to fly! At this time of year it is possible to separate adults and juveniles by the condition of their feathers. A bird with obvious gaps in its wings or an uneven tail will be an adult, part way through its annual moult. In contrast the plumage of recently fledged young Red Kites is usually far more neat and tidy, with all feathers recently grown and of the same age.

Moulted Red Kite feathers can often be found on the ground in areas where the bird is common. In early summer they are often found in woodland close to an active nest site. Later in the summer they may be found more frequently over any area of countryside frequented by kites. Red Kite fieldworkers may find it hard to resist the temptation to pick them up, and can build up quite a collection over the years.

A LULL IN ACTIVITY

Keen observers of the Red Kite will notice a clear drop-off in activity in late summer and early autumn. Even areas where the bird is very common and almost ever-present in the skies at other times of year may seem virtually devoid of kites at this time. This is partly because, with the end of the breeding season, there is no more of the endless ferrying of food items back to the nest site as adults provision their young. Activity levels are also affected by the rapid moult that most of the adult population is undergoing at this time of year, which makes flying more onerous and tends to result in birds spending less time in the air.

Late summer is a time when food tends to be abundant, as the countryside is full of inexperienced and vulnerable young birds and animals born in the summer, not all of which will survive into the winter. As a result, Red Kites may not need to make extended foraging flights in order to locate animal carrion and keep themselves well fed. Food may be found quickly early in the morning, meaning that much of the rest of the day can be given over to perching quietly and unobtrusively in a tree. In winter, perched birds tend to be quite obvious, standing out against the branches in the bare, leafless landscape, but in late summer, with the leaves still on the trees, they are often far less noticeable.

SEPTEMBER

WANDERLUST

IN BRITAIN, ONCE A RED KITE HAS PAIRED UP and bred for the first time it tends to remain faithful to the same area, remaining within a few kilometres of its nest site throughout the year, often for the rest of its life. Many younger birds, however, have more of an urge to explore, and it is by no means uncommon for them to make long-distance flights away from the area where they were born. These are often initiated in the autumn, soon after they have become fully independent, and by the end of September many young birds will already be settled in the area where they are going to spend the coming winter. Some birds wait until the following spring before making exploratory movements. Both sexes can disperse, but studies of tagged birds have shown that a higher proportion of females than males undertake long-distance movements.

In England and Wales, young birds disperse in an apparently random fashion, with movements in all directions of the compass. These wandering birds can turn up just about anywhere in Britain, although the bird's social nature means that they often end up in areas with an existing Red Kite population. When they are old enough to make their first breeding attempt many individuals return to the place where they were born though some remain with Red Kites in other populations, almost certainly never to return to their area of birth.

Radio-tracking has shown that a few youngsters struggle to make up their minds exactly where they wish to be, as with a bird released in Yorkshire in late summer as part of the reintroduction programme. The following spring this individual was recorded 250 kilometres away from its release site, associating with Red Kites in the Chilterns, before being found back in Yorkshire about two weeks later. Not content with this, it then made a further trip to the Chilterns later in spring, only to return, once again, to Yorkshire. Such behaviour may not be especially unusual – but it has only been revealed recently by the use of modern research methods such as radio-tracking.

Many young birds from the Scottish Red Kite populations move south and west in the autumn, and there is a small, but regular, passage of birds across the sea to Ireland. At least two Scottish birds have been recorded as far south as Iberia. However, top prize for the most unexpected movement goes to another Scottish kite that fledged from its nest in northern Scotland in July and was then found over 1,000 kilometres to the northwest in Iceland in December – representing the first ever record of a Red Kite for that country. It survived two harsh Icelandic winters before being found in poor condition covered in Fulmar oil. Fulmars spit this foul-smelling oil as a defence mechanism when threatened close to their nest site – and very effective it is too, as many seabird researchers will testify. Although the bird was returned to Scotland for rehabilitation there was, unfortunately, no happy ending to the story. Just six weeks after being re-released in Scotland it was recovered dead.

THE RED KITE AS A MIGRANT

In contrast to British birds, Red Kites in parts of northern and central Europe are truly migratory, with most birds (both adults and young) undertaking a long-distance movement each spring and autumn, to and from the breeding areas. Although the Red Kite is a fairly hardy bird, able to withstand the cold winter conditions that sometimes prevail in Britain, it is vulnerable to the very harsh and inhospitable conditions found frequently in winter further north and east. Prolonged snow cover is likely to pose a particular problem, as it makes detecting food sources on the ground far more difficult. The majority of birds in these populations spend the winter in Iberia, where conditions are far more favourable, and large numbers move through mountain passes in the Pyrenees between France and Spain in spring and autumn. The autumn passage starts in earnest towards the end of September, and many thousands of birds pass through these mountains, in groups of up to a hundred birds, during the following few weeks. A handful of Red Kites are seen at the famous raptor migration watch point in Gibraltar in most years, suggesting that at least small numbers travel on as far as North Africa to spend the winter.

In recent years, a higher proportion of birds in central and northern Europe have been remaining on their breeding grounds throughout the year. If, as seems likely, this is the result of increasingly mild winters, then the effects of global warming may further reduce the number of birds undertaking a long-distance migration in the years ahead.

Long-distance migrants may, unwittingly, stray off course from their usual migration routes, and some continental birds end up in Britain each year, particularly at coastal sites in southern and eastern England. It is known from studies of DNA taken from blood samples that Red Kites originating from mainland Europe have occasionally stayed on to breed in Britain. By the same token, it is likely that wandering birds from Britain occasionally end up breeding in populations on the continent.

THE USE OF NEW TECHNOLOGY

Following the movements of birds in detail using satellite tags has, in recent years, become possible for large species such as the Red Kite. These tags transmit data to satellites which in turn relay the information back to the ground, ultimately ending up on the computer screens of researchers studying the birds. With the latest tags, accurate daily locations are possible, allowing movements to be followed at a level of detail that was unimaginable only a few years ago. And once the tag has been fitted to the bird there is no need to even leave the office to the collect the information.

Satellite tracking has been carried out on a small number of Red Kites in parts of Europe and has confirmed the relatively leisurely nature of the autumn migration. One young bird, fitted with a tag in Switzerland, began to head southwest in the last week in September. It then spent almost three weeks on the edge of the French Pyrenees, before finally crossing the mountains into northern Spain. It reached its final wintering site in northern Spain more than a month after its journey had begun. In Scotland, a recent project utilised satellite tags on Red Kites and posted their movements online, giving everyone the opportunity to follow the fate of individual birds. It is hoped that, in the coming years, increasing use of this technique will greatly improve our understanding of the Red Kite and its patterns of movement.

New technology is increasingly being used to improve our understanding of many aspects of bird behaviour. The use of satellite tags, for example, has helped shed light on the timing of migration and the routes taken. Satellite and GPS tags have also helped to show, in detail, how birds utilise the local landscape in the breeding season and in winter, revealing which areas are exploited most often and how far birds will travel from breeding and roosting sites in order to search for food.

TO FEED OR NOT TO FEED?

THROUGHOUT HISTORY, THE RED KITE HAS DEPENDED on food unwittingly provided by humans for its survival. In urban areas it once scavenged on animal waste thrown out into the open streets. And in rural areas, dead livestock has long provided a useful food source, especially during the colder winter months when other food may be scarce, and in the spring lambing season when food is required to bring birds into breeding condition. In some countries, waste from abattoirs, food-processing plants and refuse tips provides an additional food source that kites have been able to exploit.

More than any other bird of prey in Europe, the diet of the Red Kite is associated with human activities. While this has helped to make the bird such an adaptable species and one that is able to thrive in close proximity to human settlements, it has also made it vulnerable to changes in farming practices and waste-disposal methods. In recent years the provision of food in gardens has become much more popular, though this can be controversial.

CHANGES IN LEGISLATION

The rules governing the disposal of animal waste in the European Union have changed considerably in recent times, and farmers are now required to pick up livestock carcasses so that they can be taken off the farm and disposed of safely, often through incineration. This has resulted in a significant reduction in food available for birds of prey, especially the scavenging Red Kite. Regulations governing the disposal of animal remains at processing plants and at open rubbish dumps have also been tightened, reducing the availability of another food source exploited by the Red Kite.

Up to 400 Red Kites have been attracted to the feeding station at Gigrin Farm in mid-Wales (see page 118), forming one of the most impressive wildlife spectacles that Britain has to offer. Higher numbers are seen during spells of severe weather, when the competition for food in the local landscape becomes more intense.

RED KITE FEEDING STATIONS

The deliberate provision of food for birds of prey has been used as a conservation technique for vultures and eagles in Europe and elsewhere for many years. They have recently taken on greater importance because of the tighter regulations on disposal of animal waste and can provide a valuable source of food in areas where dead livestock is now rarely available. In Scotland and Wales, a number of feeding stations aimed primarily at Red Kites have been established. Perhaps the best known is at Gigrin Farm, near Rhayader in mid-Wales. At this, and other sites, food is put out at the same time each day throughout the year. Up to 350 kilograms of meat is used each week at Gigrin Farm, all of which must be fit for human consumption in order to comply with the strict regulations.

Unlike the whole carcasses used at feeding stations for vultures in southern Europe, the food provided for Red Kites is in the form of small pieces of meat. This reflects the very different feeding behaviour of Red Kites. Whereas vultures

Pieces of meat snatched from the ground at feeding stations may change hands many times, amid a mêlée of squabbling birds, before they are finally consumed. It is little wonder that Red Kites seek to carry food away from the feeding site rather attempting to feed on the spot.

land on the ground in order to tear pieces of meat from large carcasses, Red Kites are very reluctant to do so. They much prefer to dive down and snatch a manageable amount of food from the ground so that it can be taken away to a secure perch to be consumed, well away from the feeding frenzy where there is a constant risk of food being stolen. Thanks to feeding sites such as Gigrin Farm, and Bellymack Hill Farm in Dumfries and Galloway, thousands of people have now taken the opportunity to watch Red Kites, and other species such as Buzzards and Ravens, at close range from purpose-built hides.

FEEDING IN GARDENS

In some of the areas where they have been reintroduced in England and Scotland, people have increasingly taken to providing food for the local Red Kites in their gardens. This has become commonplace in villages and even towns and cities in the Chilterns, and the birds have quickly learnt about this potential food source. Groups of Red Kites can now often be seen floating over village gardens at little above rooftop height, and built-up areas frequently support higher concentrations of birds than the open countryside in between.

When food is spotted from the air, the wings are folded back and the bird dives down rapidly to snatch it up from the ground before carrying it away to a feeding perch or, in the breeding season, ferrying it back to the nest site. The Red Kite's relative lack of fear of people and supreme aerial agility mean that

Householders in the Chilterns have had great success in attracting Red Kites to village gardens. The birds do not usually linger, preferring to snatch up pieces of meat in flight for consumption on the wing or in a nearby tree. Feeding has become contentious in parts of the Chilterns, but provided that a few simple guidelines are followed there is no reason why it should not be undertaken – and it certainly brings great pleasure to many people.

it will take food from even very small gardens, providing spectacular, close-up views from the comfort of the living room. Local pubs, cafés and garden centres have taken to feeding Red Kites as an added attraction for patrons, and butchers have even started selling unwanted offcuts of meat as 'kite-scraps'.

Not everyone believes that feeding Red Kites in gardens is a good idea. Some people see it as unnatural human interference, resulting in artificially high concentrations of birds and reducing the rate at which they spread out into the surrounding countryside. Concerns have also been raised about the quality of some of the food provided – scraps of meat or leftovers from the Sunday roast might not be especially healthy for kites. Putting out large amounts of food can also annoy neighbours, as it can attract species that people see as undesirable, including rats and crows. There have been incidents where scraps of fatty meat have been picked up by Red Kites only to be unwittingly dropped on the luxury car parked in next door's drive.

There are even occasional reports from the Chilterns of people feeling threatened by Red Kites. Children at one local primary school apparently felt concerned about losing items from their packed lunches to birds circling low overhead and diving down to investigate potential food sources within the confines of the playground. One child was scratched on the hand by the talons of a bird seeking to snatch food, and several people have expressed concern that small pets may not be completely safe from such a large and bold bird of prey.

The Red Kite has always had a close link with human settlements and has been taking advantage of food scraps resulting from our activities from medieval times onwards. Historical descriptions include the occasional account of food being snatched from the unwary – and this behaviour is certainly not beyond the closely related Black Kite in parts of its range in Asia and Africa. However,

A study by the University of Reading attempted to assess the scale of feeding in the town. Red Kites are regularly seen over urban Reading during the day though there are few, if any, breeding or roosting sites. The researchers speculated that birds travelled into the town each day in order to take advantage of the food provided by people in gardens. Road transect surveys showed that between 140 and 440 birds visited the town each day, and face-to-face interviews with residents revealed that over 4% (equivalent to over 4,000 householders) put food out for Red Kites. The amount of food provided was sufficient to explain the numbers of birds visiting the town. Other towns and cities are now starting to attract Red Kites, and small numbers of birds have even been seen over central London. It will be fascinating to see whether they are able to fully recolonise our urban landscapes by nesting and roosting in large gardens or parks, or whether they will continue to fly in for food from the surrounding countryside.

provided that some basic advice is followed (see below), putting food out in gardens is unlikely to cause significant problems. It is perhaps little different to putting peanuts out for tits and other garden birds, which is now such a common occurrence and brings pleasure to huge numbers of people.

In areas such as mid-Wales, where natural foods can be hard to come by, organised feeding is likely to be a substantial benefit to the local Red Kite population. Far from reducing the rate of spread to new areas, feeding may help to improve survival rates and so result in a larger pool of birds being available to help with recolonisation of currently unoccupied countryside.

In parts of their range Red Kites can be rather scarce in intensively managed areas of the countryside. They may be found in greater numbers over towns and villages, where food is often easier to come by.

Guidelines for feeding Red Kites in gardens

- Artificial food based on meat that has been processed for human consumption may contain harmful additives such as salt and should be avoided or restricted to very small quantities.

- Food derived from complete animal carcasses, including skin and bone, is a healthier option for Red Kites as it contains beneficial nutrients and minerals. Whole mice, which can be bought from pet-food suppliers, are ideal.

- In lowland areas, where natural food is generally abundant, only small amounts of food should be provided to avoid birds becoming too dependent on handouts and to reduce the risk of attracting unwelcome scavengers such as rats. Any excess food should be cleared away at the end of each feeding session.

- Neighbours should be consulted and their views considered before food is provided. Not everyone is a fan of large birds of prey, and people may have concerns about their children or pets being frightened, or fear that other desirable species may be scared away by large numbers of Red Kites.

- Red Kites are large enough to pose a threat to aircraft, and there have been a number of collisions. Feeding should not be undertaken close to active airfields. Staff at several airfields in England have been issued with licences to allow problem birds to be shot, and feeding nearby might increase the need for such licensed control in future.

OCTOBER

THE COMMUNAL ROOST

THE HIGHLY SOCIAL NATURE OF THE RED KITE is revealed to its full extent from late autumn onwards as large communal roosts begin to form in areas where they are common. These support the highest numbers of birds in the middle of winter (see *December*) but, by October, they are already starting to build up, and each night more and more birds head for the same areas of woodland at the end of the day's foraging. Some traditional communal roost sites are used repeatedly each winter over many years and may support a hundred or more birds. Young birds presumably learn about these sites by copying the behaviour of older birds that have used them in previous years.

In the newly established reintroduced populations, the majority of birds initially used a single site for roosting, and almost the entire population could be perched together in the same group of trees each night. As populations expanded, the number of roost sites increased and the average number of birds attending each roost declined, reflecting the greater choice available.

Each bird will often use several different sites during the course of the winter, and there have even been records of individuals switching between nearby roosts in the middle of the night. As with many other aspects of Red Kite behaviour, this was first revealed by radio-tracking. Ingenious automatic computer loggers were set up by members of the Southern England Kite Group to record the radio signals of tagged birds at communal roosts in the Chilterns. Researchers checking the data found that the signal from a tagged bird disappeared from one roost and then, a few minutes later, was picked up at another, a few kilometres away. Quite why birds would risk flying around in total darkness in order to switch roosts is something of a mystery, although perhaps it is the result of unexpected disturbance at the initial roost.

At dusk ... pre-roost ... we watch numbers as they make their way along the Meon valley to the roost. Why here? Disturbance at other sites perhaps. Or just natural instinct, mirroring choices that kites would have made hundreds of years ago when they were last common in the area.

ATTENDANCE PATTERNS

Young birds tend to use communal roosts every night from the autumn through until the following spring. This has been confirmed by reading the wing-tags of birds arriving at roosts and by studies of radio-tagged birds whose presence in a roost can be confirmed very easily, even after dark when they cannot be seen. Adult birds also attend communal roosts, although not with the same regularity as young birds. Adults with an established breeding site will frequently roost together as a pair close to their nest, even in the depths of winter when thoughts of breeding must be very far away.

In southern Spain, there is a clear contrast in behaviour between resident birds and birds from central and northern Europe that visit for the winter. The resident breeders remain together as pairs and spend the night on, or close to, their nest, whereas the wintering birds join with others to roost communally. Young birds from the resident population which have yet to pair up and establish breeding territories often join the communally roosting wintering birds. There is also a difference between these two groups of birds during the day. The resident adults tend to forage on their own within a relatively small area centred on their nest site, while the wintering birds usually forage in loose groups and range over a far wider area in search of food.

THE DAILY ROUTINE

Although visitors to a major communal roost site are likely to see at least small numbers of Red Kites at any time of day, the influx of birds usually begins in earnest around mid-afternoon, with birds then arriving in increasing numbers as the light starts to fade and the day draws to a close. It is likely that individuals that have had a successful day's foraging return first, and it is sometimes possible to see the swollen crops on birds that have recently eaten a decent meal as they pass low overhead. Those that arrive later in the day may not have been so lucky and so have continued foraging for longer in order to try to find food. A few stragglers may even arrive when the light has almost completely gone and when most birds are already safely tucked away within the roost wood.

The pattern of leaving the roost in the morning is similar, with small numbers (perhaps the hungriest), leaving soon after first light and then a steady flow of birds leaving as singles and small groups over the next hour or two. There tends to be far less activity around the roost area in the mornings than in the afternoons (see *November*), presumably because most birds are keen to get on with the essential business of searching for food. Red Kites will forage at any time of the day, but observations suggest that there is often a peak of activity in the first few hours after daybreak. Birds that find a substantial meal in the morning can afford to take it easy for the rest of the day and conserve their energy.

If the weather conditions are poor for foraging, with dense fog or heavy rain, for example, then some birds may remain in the roost area for the whole day. The Red Kite is a large enough bird to be able to forgo food for the occasional day. In very poor conditions, it no doubt makes more sense for a well-fed bird to conserve energy and remain at the roost, rather than risk expending valuable energy on a period of foraging when success is far from guaranteed.

An early afternoon recce of the roost site on a dull, drizzly day and surprise, surprise the birds were already in. Or perhaps they had never left, deciding it was better to remain local than waste energy searching the valley for food in such unpromising conditions.

NOVEMBER

THE BENEFITS OF COMMUNAL ROOSTING

RED KITE ROOSTS CONTINUE TO BUILD UP through the autumn and during the early part of the winter. Although only a minority of birds of prey gather together to roost in large numbers, communal roosting is a widespread habit among many different bird species. Some small birds, such as Long-tailed Tits and Wrens, roost together during cold winter nights primarily to conserve body heat. Another benefit of roosting in numbers for small birds comes from a reduced risk of predation through having many eyes in the same area on the lookout for potential threats. For the Red Kite, neither of these explanations seems satisfactory. Observations at roosts show that Red Kites do not spend the night huddled closely together, even in the coldest conditions, and so keeping warm does not seen to be an important factor. And although full-grown Red Kites may occasionally be taken by large avian predators such as Eagle Owls and the larger eagles, this is uncommon, and certainly not a significant threat for birds in Britain.

Researchers have now shown that birds such as the Raven and American Black Vulture roost communally in order to improve their ability to find food during the following day. The birds do not actively exchange information about the location of food, as was once thought, but unsuccessful or inexperienced birds can learn about food sources by following more knowledgeable birds from the roost in the morning. These species often feed on medium-sized or large animal carcasses that may last for several days and are sufficient to feed a number of birds – and the same explanation may well hold for the Red Kite.

It is not known for sure whether Red Kites actively follow each other in order to track down specific food items, but it is undoubtedly true that Red Kites

So-called 'network foraging' in action. Red Kites often search for food in loose groups, keeping an eye on each other as much as on the ground below. If one bird drops down to a likely food source the others will be alerted to the opportunity and quickly head for the same spot. Red Kites have superb eyesight and can probably keep tabs on the behaviour of other birds over a considerable distance. This helps to explain what sometimes happens when food is put out in gardens. At first there may be no response but, sooner or later, the food will be spotted. Perhaps just a single bird will make the first move to investigate, but it will quickly be joined by several more – and within a few minutes there may be 15 or 20 birds circling over the garden.

forage most effectively when they are part of a loose group. Birds on the lookout for food are not only scanning the ground below for animal carcasses but are also keeping an eye on other Red Kites in the area. If one bird sees something and lands to investigate, then the others quickly converge on the same spot. As animal carcasses are often large enough to satisfy several birds, there is no disadvantage to the first bird in having to share its find. This is known in the scientific jargon as 'network foraging', and birds no doubt join communal roosts to ensure that they will be able to make best use of this technique on the following day. Birds roosting in isolation must rely on their own abilities in order to find food. This may not be a problem for experienced adults roosting on nesting territories with which they are very familiar, but it could leave young and inexperienced birds far more vulnerable to the threat of starvation.

Aerial interactions are often observed at communal roost sites, mainly involving groups of young birds. Most activity tends to be on windy days when the birds can exploit the conditions and stay in the air for long periods with a minimum of effort. Here several juveniles are chasing each other. It may appear to the casual observer

YOUNG BIRDS TUMBLING PLAYING
AT THE ROOST.   2014

that birds interacting in this way are simply enjoying themselves – and it seems clear that there is an element of 'play' involved. At the same time it no doubt serves a greater purpose, honing difficult flight skills and perhaps impressing birds of the opposite sex, including potential future mates.

SOCIAL INTERACTIONS AND PLAY AT ROOSTS

One of the great joys of visiting a Red Kite roost in winter (see *December*) is the opportunity to watch the frequent aerial interactions between birds before they finally settle down for the night within the roost wood. Such behaviour is most apparent in windy weather when the birds make best use of the conditions to fly around the roost area with a minimum of effort. Young birds can often be seen chasing each other, and sometimes indulging in spectacular play-fights, diving at each other with talons outstretched as if about to launch a genuine attack, only to pull away at the last second. This sort of activity can be rather surprising if you are familiar with the Red Kite's more typical, languid flight as it drifts slowly across open country in search of food. It is not entirely clear what is behind such interactions. In many cases, the birds are probably just playing; filling the time before the end of the day and using the opportunity to hone aerial skills that may be useful later in life when they are paired up and have a nesting site to defend.

Another form of play seen at roosts involves birds diving down to tree-top level to snatch at small branches and dead leaves with their talons. If successful, they may then repeatedly drop and re-catch the snatched material before finally letting it fall to the ground. Birds have also been watched performing a similar type of behaviour close to ground level, snatching up leaves from a growing crop. It is easy to imagine that this sort of activity could improve coordination skills – and it has a direct parallel with behaviours that will be useful in future when collecting nest-building material. We tend to be most familiar with examples of play in the mammal world – for example Fox cubs rough-and-tumbling outside their earth – but there is little doubt that it also serves a very useful purpose in young birds.

Given the large numbers of birds that are together in a small area around a roost site, it is inevitable that unpaired youngsters will encounter many other birds, including potential future mates. Some of the aerial interactions seen at roosts may involve birds assessing each other as the first stage in the pairing process. Red Kites usually pair for life so, as with humans, the selection of a suitable partner is something that is clearly worth taking a little time over. It would be fascinating to know how many established pairs first encountered each other in a mêlée of birds flying over a communal winter roost site.

... just when we thought the day had no more to give ...

THE ROOSTTIME BALLET. 2016 DP
AT THE M-VALLEY ROOST

DECEMBER

A WILDLIFE SPECTACLE

TOWARDS THE END OF THE YEAR, with the days at their shortest and frequent spells of cool and damp weather providing less than ideal foraging conditions, Red Kites must make the most of every opportunity to find food. The advantages of foraging in loose groups (see *November*), are more important now than at any other time and, as a result, communal roosts tend to support the highest numbers of birds at this time of year.

There are few better ways to enliven an otherwise dreary December afternoon than to take in the spectacle of Red Kites massing together as they gather at a roost site. In the absence of specific information, the best bet is to choose an area where Red Kites are often seen during the day and watch from mid-afternoon from a high vantage point to see if a concentration of birds can be detected. It is unusual for a Red Kite to be more than a few kilometres from its roost site during the day, and so following the flight lines of birds from mid-afternoon onwards can be a useful means of locating roosts. Immature birds, heading purposefully in a straight line late in the day, deserve particular attention. Depending on the nature of the landscape and the availability of good vantage points, it may take a number of days to track down a roost using this method, but the task is an enjoyable one and the effort well worthwhile.

The following passage is taken from our earlier book *The Red Kite*, a detailed overview of the ecology and status of the species. It describes the build-up of birds at a typical large communal roost site in England in the depths of winter:

> *During the latter part of the afternoon a slow trickle of birds*
> *gradually increases until single birds and small groups are*

arriving almost constantly from all directions. Initially, birds can be seen over a wide area but they are increasingly drawn together, either perching in groups in prominent trees along a hedgerow or at the edge of the roost wood, or wheeling and chasing together over the roost area. New arrivals are frequently greeted by calling from birds that are already present. This, together with the aerial activity, may help to draw more birds towards the roost.

In the same chapter is the following description – a clear example of finding oneself in the right place at the right time:

On one memorable clear and still early evening at a roost site in southern England, just as the sun was finally disappearing behind the low Chiltern Hills, a loud clattering noise from a nearby farmyard disturbed upwards of 100 Red Kites from their perches within a small wood in the valley below. They rose above the wood with a heavy, flapping flight, initially in a tight group, but gradually fanning out over a wider area of the surrounding countryside, before returning, in small groups, to resettle in trees close to the original roost.

The site described above is regularly monitored by members of the Southern England Kite Group, and a maximum of 217 birds have been counted here in a single afternoon. Roosts in Spain, where most continental birds spend the winter, can support a staggering 500 birds – and as populations continue to rise rapidly in Britain perhaps roosts of this size will be seen here in the future. Gatherings at feeding stations in Britain already sometimes involve several hundred birds. Those who have visited communal roosts are certainly left in no doubt that they provide one of our most impressive wildlife spectacles and, thankfully, one that is becoming increasingly common in Britain.

THE INFLUENCE OF WEATHER ON ROOSTING BEHAVIOUR

Those keen to witness spectacular aerial gatherings of Red Kites, as they wheel and chase above the roost wood, would be well advised to pick a windy afternoon for their visit. By contrast, when I was monitoring Red Kites in the recently reintroduced population in Northamptonshire I much preferred the

On calm days groups of Red Kites often perch prominently in isolated hedgerow trees or on the edge of a wood when they first arrive in the roost area (like ripe fruit on a tree, according to one seasoned observer!). As the light starts to fade they fly in ones and twos into the woodland trees where they will spend the night.

calmest days for roost watching, for a reason that will become clear. Without a wind to exploit, there is much less aerial activity and early arrivals tend to perch quietly on the edge of the roost wood or in isolated trees in nearby fields before they move to the roost itself later in the afternoon. Birds clearly like to perch close to each other, and once the first arrival has chosen a particular tree, seemingly at random, other birds follow suit until there may be 20 or more together in the same tree, occupying almost every available perch. If perching space becomes scarce then birds will attempt to displace one another as they come in to land. During this 'pre-roosting' period, birds spend time preening to keep their plumage in good condition and can sometimes be seen pecking at their feet to dislodge soil, picked up from foraging in muddy fields.

And the reason why Red Kite fieldworkers prefer calm days? Is it the lack of a cold winter wind biting through several layers of clothing that most appeals? While that may indeed be a factor, the main reason is that calm days encourage larger numbers of birds to perch in the open when they arrive in the roost area,

There are few more spectacular sights in the British landscape than a large group of Red Kites twisting and turning in the air above their communal roost site on a windy midwinter afternoon.

DAN POWELL.

They drift towards the roost in ones and twos. Some going straight in, others making a circuit of the field, perhaps seeking an evening snack. Then it's over – and the last views before the light finally fades are of a series of silhouetted pint-pots in the trees. The memorable evenings are those when, on a whim, or perhaps disturbed, the birds sail out as a flying phalanx, sometimes low over our heads, before gently and gracefully returning to the trees.

A SKY FULL OF KITES IN HAMPSHIRE DEC 2014
D.T.

which provides an ideal opportunity to read wing-tags and so gather useful information on the birds in the local population. It is also far more straight-forward to estimate the total number of birds using a roost site on windless days when they tend to fly straight into the wood from their pre-roost perch. On windy days when there is more aerial activity it is very difficult to keep track of individuals, and counting the overall number of birds present becomes almost impossible.

WORLD STATUS

ALTHOUGH THE OCCASIONAL BIRD MAY STILL BE FOUND in parts of North Africa and the Middle East, the Red Kite is overwhelmingly a European bird. As such, it has a far smaller world range than its close relative, the Black Kite, which is a familiar sight across large parts of Europe, Asia and Africa. Following the last ice age, the Red Kite would have been one of many birds to benefit from an increasing human population in Europe, and the associated clearing of woodland for farming and human settlements. This is because, although it requires trees for nesting and roosting sites, it is very much a bird of open country, and this is where it finds almost all of its food.

A rapidly increasing human population also brought its problems, and in continental Europe, as in Britain, persecution has had a huge impact on the bird's current distribution. In areas where it is left alone, the adaptable Red Kite is still common, but where it has been targeted by persecution or is killed accidentally by poison baits intended for another species, it is much reduced or has even been wiped out entirely. A recent estimate suggested that in the seventeenth century the Red Kite would perhaps have been breeding across up to 8 million square kilometres in Europe. It now occupies little over 1 million square kilometres.

Red Kite numbers had become so low by the second half of the twentieth century that there were real concerns the species could continue to decline towards eventual extinction. In the 1980s, the estimated world population stood at just 5,500–15,000 pairs, although this may have been something of an underestimate based on incomplete information in parts of its range.

The total world Red Kite population currently stands at an estimated 23,000–29,000 pairs – but to put this in perspective, consider the more familiar

Common Buzzard. A recent assessment suggested that there were 903,000 pairs of Buzzards in Europe. And, following a sustained recovery from past persecution, there are now thought be over 60,000 pairs in Britain alone, more than double the total world population of the Red Kite. The following sections provide more details for European countries that still support substantial populations of Red Kites, along with a brief summary of recent changes in status in each area.

NORTHERN EUROPE

The Red Kite breeds no further north in Europe than northern Scotland (where it has been reintroduced), Denmark and the southernmost parts of Sweden. A few pairs once bred in southeast Norway but these were lost as a result of human persecution. The Swedish population has made a substantial recovery in the last few decades and now stands at close to 2,000 pairs. Denmark was recolonised from Sweden in the 1970s and now supports over 100 pairs. By the late 1980s, Sweden supported enough Red Kites to allow small numbers of young to be taken from nests each year to support the reintroduction programme in Scotland. Many of the Scandinavian birds migrate to Spain for the winter, but some now remain all year, perhaps due to the increasingly mild winters.

NORTHWEST EUROPE

Despite recent declines in parts of the country, France still supports around 2,000–3,000 breeding pairs, mostly in the south and east. Poisoning remains a problem either through the deliberate use of poison baits to kill predators or modern, highly toxic rodenticides used to control vole populations on agricultural land. A reduction in the number of open refuse tips has also been suggested as a possible factor leading to declines.

Belgium was recolonised in the 1970s and, following a steady increase, now supports up to 100 pairs. Neighbouring Luxembourg has a similar population and, as in Belgium, numbers are increasing. The Netherlands is regularly visited by Red Kites, and it is hoped that a breeding population may become firmly re-established in the country in the coming years.

CENTRAL AND EASTERN EUROPE

Germany supports more than three times as many breeding Red Kites as any other country, despite declines in some areas during recent years. The population was recently estimated at between 9,000 and 14,000 pairs, close to half of the total world population. Declines have been blamed on the intensification of farming, which may have reduced the bird's food supply, and also on losses from human persecution in Spain, where most of the German birds spend the winter. This has been confirmed by the fact that some poisoned birds in Spain have been found wearing leg-rings fitted to them as nestlings in Germany. The increase in wind farms in Germany may also present a problem, as Red Kites are known to be regular victims from collision with these structures. Germany supplied many of the young birds for release in Scotland during the later years of the reintroduction programme.

The Swiss Red Kite population has increased rapidly since the 1970s, and this small country now supports over 1,000 pairs. Neighbouring Austria supports a small number of pairs, following natural recolonisation, although persecution remains a serious problem. The Czech Republic, like Austria, has been recolonised naturally following extinction due to human persecution, and more than 150 pairs are now present.

Red Kites in Poland are found mainly in the north and west, where the population is estimated at around 1,000 pairs. Other eastern European countries, including countries of the former Soviet Union, have either lost their Red Kites as a result of human persecution or support only a small remnant population. It is hoped that here, as in other areas, changing attitudes towards large birds of prey will allow the species to make a comeback.

As with the Swedish birds, some breeding birds in central and eastern Europe now remain in the breeding areas all year round, and this trend may well increase if climate change leads to a reduction in the severity of winters.

Southern Europe

Spain is by far the most important country for the Red Kite, not because of its breeding population, but because many of the birds that breed in central and northern Europe move south across the Pyrenees to winter here. A recent survey found no fewer than 29,000–30,000 birds wintering in Spain (in addition to the resident Spanish birds), although this represents a substantial decline from the 54,000–62,000 birds thought to have been present ten years previously. Part of this decrease may reflect the greater numbers of birds remaining on their central and eastern European breeding grounds for the winter rather than migrating south. Losses from poison baits are a concern, with hunters blaming birds of prey for declines in game species and taking the law into their own hands. Problems in Spain are especially worrying, because the loss of wintering birds has the potential to affect the important breeding populations to the north and east.

It is thought that just 2,000–2,200 breeding pairs remain in Spain, representing an alarming decline of almost 50% in just ten years. The province of Segovia in central Spain provided many of the birds for release in Northamptonshire in the English Midlands as part of the reintroduction programme but has since seen a dramatic slump of 80–90% in the local population. In the future, it is not inconceivable that we may need to return the favour and take birds from the reintroduced populations in Britain for release in Spain.

In Portugal, the Red Kite is still heavily persecuted, and fewer than 100 pairs are believed to remain. Italy is the only country in southern Europe that has seen a recent increase in breeding Red Kites, despite continued persecution, aided by a reintroduction programme in Tuscany. There are now around 300–400 pairs.

The Red Kite is an adaptable bird and is still found in a wide range of landscapes across Europe, despite problems with persecution in some countries. In the south, as here in southern Spain, it shares the skies with fellow scavengers including the Black Kite and Griffon Vulture and enjoys long, baking hot summers. In the north and east of the continent it may be seen with White-tailed Eagles, and it copes well with summers that are much cooler and wetter.

Of the five major Mediterranean islands within the Red Kite's range, only Corsica has seen a population increase. Between 200 and 300 pairs are thought to breed there. The two Spanish islands of Majorca and Minorca once supported substantial populations but both have suffered major declines, and very few birds now remain on either of these islands.

Countries supporting the largest Red Kite breeding populations

Country	Recent estimates of breeding population (pairs)	Trend
Germany	9,000–14,000	Decrease/stable
Britain	5,000+	Rapid increase
France	2,000–3,000	Decrease
Spain	2,000–2,200	Decrease
Switzerland	1,000+	Increase
Sweden	~ 2,000	Increase
Poland	~ 1,000	Increase

Countries recently recolonised by Red Kites

Country	Year recolonised	Current population (pairs)
Austria	1970s	< 100
Belgium	1973	< 100
Czech Republic	1970s/80s	150+
Denmark	1970s	100+
England	From reintroduction project starting in 1989	Estimated 4,000+
Scotland	From reintroduction project starting in 1989	300+
Ireland	From reintroduction projects starting in 2007 (Republic of Ireland) and 2008 (Northern Ireland)	85 pairs in the Republic, 20 pairs in Northern Ireland (2017)

THE FUTURE

The Red Kite is a highly adaptable bird and is able to thrive in a wide range of landscapes across Europe, provided it is left unmolested by humans. However, its scavenging habits and relative lack of fear of people make it highly vulnerable to persecution, especially poisoning, and it is often the first bird of prey to be lost from an area where poison baits are in regular use.

As we have seen, the Red Kite has had mixed fortunes in different parts of its European range in recent times. There has been a very welcome recovery in some countries where there is now less human persecution, and the bird has been able to recolonise several countries from which it was wiped out in the past. But there have also been some worrying declines, including in several of the countries that support the highest numbers of birds and so are especially important. The overall world population of 23,000–29,000 pairs is certainly very far below the level that would be expected in the absence of human impacts – and, sadly, the bird is no longer present across large areas that appear to provide perfectly suitable habitat.

As discussed earlier, if the densities of birds currently present close to the reintroduction sites in southern and central England were to be replicated across all suitable areas of countryside then the total population in Britain alone could, in future, exceed 50,000 pairs, around double the current world population. Already, Britain probably supports more breeding pairs than any other country except Germany, though the lack of a recent full survey means we do not have an accurate figure for the total population. If urban areas can be recolonised, as seems increasingly likely, this will provide further opportunities for this adaptable species.

In the years ahead, the fortunes of our most graceful and elegant bird of prey will tell us much about our ability and willingness to coexist with wildlife. We are quick to point the finger at other countries with a record of killing protected birds, for example in the long-term campaign against the illegal killing of migrant songbirds and birds of prey in Malta. Perhaps we should spend more time looking closer to home, particularly to the continued illegal persecution of birds of prey in upland areas where grouse shooting is the predominant activity. For a stark reminder of the scale of the impacts of illegal persecution, a comparison between the status of the Red Kite in southern England and northern Scotland is all that is required.

Despite the success of conservation efforts in Britain, and encouraging signs that reintroduction in Ireland will be successful, the future of the Red Kite is far from completely secure. Populations are in decline in some areas of Europe and the species remains a rare sight in large parts of its potential range, including across much of Britain and Ireland. It will only make a full recovery if the problems associated with illegal persecution and accidental deaths from poisoning and other human impacts can be tackled effectively. If that can be done, we can look forward to the Red Kite once again becoming the common and widespread bird that it would have been in the past – a familiar and well-loved species available for all to see and enjoy.

FURTHER READING

The following list is by no means a comprehensive review of the literature on Red Kites, but it does contain the key sources of information used during the preparation of this book. It includes a mixture of books, reports and scientific papers, all of which should be readily available through booksellers or, in the case of scientific papers, through academic libraries. The summaries of many of the papers listed are available for free online through search engines such as *Google Scholar*, although payment may be required to access the full texts.

Aebischer, A. (2009) *Der Rotmilan: ein faszinierender Greifvogel*. Haupt Verlag, Bern. (In German but with a superb collection of colour photographs.)

Balmer, D.E., Gillings, S., Caffrey, B.J., Swann, R.L., Downie, I.S. & Fuller, R.J. (2013) *Bird Atlas 2007–11: The Breeding and Wintering Birds of Britain and Ireland*. BTO Books, Thetford.

Brown, A. & Grice, P. (2005) *Birds in England*. Poyser, London.

Bustamante, J. (1993) Post-fledging dependence period and development of flight and hunting behaviour in the Red Kite *Milvus milvus*. *Bird Study* 40: 181–188.

Bustamante, J. & Hiraldo, F. (1990) Adoptions of fledglings by Black and Red Kites. *Animal Behaviour* 39: 804–806.

Carter, I. (2007) *The Red Kite* (2nd edition). Arlequin Press, Shrewsbury.

Carter, I. & Burn, A. (2000) Problems with rodenticides: the threat to Red Kites and other wildlife. *British Wildlife* 11: 192–197.

Carter, I. & Grice, P. (2000) Studies of re-established Red Kites in England. *British Birds* 93: 304–322.

Carter, I. & Whitlow, G. (2005) *Red Kites in the Chilterns* (2nd edition). English Nature/Chilterns Conservation Board, Princes Risborough.

Carter, I., Cross, A.V., Douse, A., Duffy, K., Etheridge, B., Grice, P.V., Newbery, P., Orr-Ewing, D.C., O'Toole, L., Simpson, D. & Snell, N. (2003) Re-introduction and conservation of the Red Kite *Milvus milvus* in Britain: Current threats and prospects for future range expansion. In Thompson, D.B.A., Redpath, S.M., Fielding, A.H., Marquiss, M. & Galbraith, C.A. (eds.) *Birds of Prey in a Changing Environment*. Scottish Natural Heritage, Edinburgh, pp. 407–416.

Cocker, M. & Mabey, R. (2005) *Birds Britannica*. Chatto & Windus, London.

Cramp, S. & Simmons, K.E.L. (eds.) (1980) *The Birds of the Western Palearctic, Volume 2*. Oxford University Press, Oxford.

Cross, A.V. & Davis, P.E. (2005) *The Red Kite of Wales* (revised edition). The Welsh Kite Trust, Llandrindod Wells.

Davis, P. (1993) The Red Kite in Wales: setting the record straight. *British Birds* 86: 295–298.

Davis, P.E. & Davis, J.E. (1981) The food of the Red Kite in Wales. *Bird Study* 28: 33–44.

Davis, P.E. & Newton, I. (1981) Population and breeding of Red Kites in Wales over a 30-year period. *Journal of Animal Ecology* 50: 759–772.

Davis, P., Cross, A.V. & Davis, J. (2001) Movement, settlement, breeding and survival of Red Kites *Milvus milvus* marked in Wales. *Welsh Birds* 3: 18–43.

Evans, I.M. & Pienkowski, M.W. (1991) World status of the Red Kite: a background to the experimental reintroduction to England and Scotland. *British Birds* 84: 171–187.

Evans, I.M., Dennis, R.H., Orr-Ewing, D.C., Kjellén, N., Andersson, P.-O., Sylvén, M., Senosiain, A. & Carbo, F.C. (1997) The re-establishment of Red Kite breeding populations in Scotland and England. *British Birds* 90: 123–138.

Evans, I.M., Cordero, P.J. & Parkin, D.T. (1998) Successful breeding at one year of age by Red Kites *Milvus milvus* in southern England. *Ibis* 140: 53–57.

Evans, I.M., Summers, R.W., O'Toole, L., Orr-Ewing, D.C., Evans, R., Snell, N. & Smith, J. (1999) Evaluating the success of translocating Red Kites *Milvus milvus* to the UK. *Bird Study* 46: 129–144.

Ferguson-Lees, J. & Christie, D.A. (2001) *Raptors of the World*. Christopher Helm, London.

Forsman, D. (1999) *The Raptors of Europe and the Middle East: a Handbook of Field Identification*. Poyser, London.

Hardy, J., Crick, H.Q.P., Wernham, C.V., Riley, H.T., Etheridge, B. & Thompson, D.B.A. (2009) *Raptors: a Field Guide for Surveys and Monitoring* (2nd edition). The Stationery Office, Edinburgh.

Hatcher, P. & Battey, N. (2011) *Biological Diversity: Exploiters and Exploited*. Wiley-Blackwell, Chichester. (An undergraduate text book with an accessible and readable chapter on the Red Kite – one of the 'exploiters', though it could equally have been included as an example of the 'exploited'.)

Holloway, S. (1996) *The Historical Atlas of Breeding Birds in Britain and Ireland: 1875–1900*. Poyser, London.

Lovegrove, R. (1990) *The Kite's Tale: The story of the Red Kite in Wales*. RSPB, Sandy. (Out of print but widely available second-hand.)

Lovegrove, R. (2007) *Silent Fields: The Long Decline of a Nation's Wildlife*. Oxford University Press, Oxford.

Lovegrove, R., Elliot, G. & Smith, K. (1990) The Red Kite in Britain. *RSPB Conservation Review* 4: 15–21. RSPB, Sandy.

Minns, D. & Gilbert, D. (2001) *Red Kites – Naturally Scottish*. Scottish Natural Heritage, Battleby.

Mougeot, F. (2000) Territorial intrusions and copulation patterns in Red Kites *Milvus milvus* in relation to breeding density. *Animal Behaviour* 59: 633–642.

Newton, I. (1979) *Population Ecology of Raptors*. Poyser, London.

Newton, I., Davis, P.E. & Moss, D. (1981) Distribution and breeding of Red Kites *Milvus milvus* in relation to land-use in Wales. *Journal of Applied Ecology* 18: 173–186.

Newton, I., Davis, P.E. & Davis, J.E. (1989) Age of first breeding, dispersal and survival of Red Kites *Milvus milvus* in Wales. *Ibis* 131: 16–21.

Newton, I., Davis, P.E. & Moss, D. (1994) Philopatry and population growth of Red Kites *Milvus milvus* in Wales. *Proceedings of the Royal Society of London B* 257: 317–323.

Newton, I., Davis, P.E. & Moss, D. (1996) Distribution and breeding of Red Kites *Milvus milvus* in relation to afforestation and other land-use in Wales. *Journal of Applied Ecology* 33: 210–224.

Ntampakis, D. & Carter, I. (2005) Red Kites and rodenticides: a feeding experiment. *British Birds* 98: 411–416.

Orros, M.E. & Fellowes, M.D.E. (2014) Supplementary feeding of the reintroduced Red Kite *Milvus milvus* in UK gardens. *Bird Study* 61: 260–263.

Orros, M.E. & Fellowes, M.D.E. (2015) Widespread supplementary feeding in domestic gardens explains the return of reintroduced Red Kites *Milvus milvus* to an urban area. *Ibis* 157: 230–238.

Pain, D.J., Carter, I., Sainsbury, A.W., Shore, R.F., Eden, P., Taggart, M.A., Konstantinos, S., Walker, L.A., Meharg, A.A. & Raab, A. (2007) Lead contamination and associated disease in captive and reintroduced Red Kites *Milvus milvus* in England. *Science of the Total Environment* 376: 116–127.

Sergio, F., Blas, J., Blanco, G., Tanferna, A., López, L., Lemus, J.A. & Hiraldo, F. (2011) Raptor nest decorations are a reliable threat against conspecifics. *Science* 331: 327–330.

Smart, J., Amar, A., Sim, I.M.W., Etheridge, B., Cameron, D., Christie, G. & Wilson, J.D. (2010) Illegal killing slows population recovery of a reintroduced raptor of high conservation concern: the Red Kite *Milvus milvus*. *Biological Conservation* 143: 1278–1286.

Snell, N., Dixon, W., Freeman, A., McQuaid, M. & Stevens, P. (2002) Nesting behaviour of the Red Kite in the Chilterns. *British Wildlife* 13: 117–183.

Snow, D.W. & Perrins, C.M. (1998). *The Birds of the Western Palearctic: Concise Edition, Volume 1.* Oxford University Press, Oxford.

Viñuela, J. & Villafuerte, R. (2003) Predators and Rabbits *Oryctolagus cuniculus* in Spain: A key conflict for European raptor conservation. In Thompson, D.B.A., Redpath, S.M., Fielding, A.H., Marquiss, M. & Galbraith, C.A. (eds.) *Birds of Prey in a Changing Environment.* Scottish Natural Heritage, Edinburgh, pp. 511–526.

Wernham, C.V., Toms, M.P., Marchant, J.H., Clark, J.A., Siriwardena, G.M. & Baillie, S.R. (eds.) (2002) *The Migration Atlas: Movements of the Birds of Britain and Ireland.* Poyser, London.

Wildman, L., O'Toole, L. & Summers, R.W. (1998) The diet and foraging behaviour of the Red Kite in Scotland. *Scottish Birds* 19: 134–140.

Wotton, S.R., Carter, I., Cross, A.V., Etheridge, B., Snell, N., Duffy, K., Thorpe, R. & Gregory, R.D. (2002) Breeding status of the Red Kite *Milvus milvus* in Britain in 2000. *Bird Study* 49: 278–286.

SOURCES OF FURTHER INFORMATION

The following is a selection of the many organisations involved with the conservation of Red Kites in Britain and Ireland in recent years which include information about this work on their websites. Links to websites can quickly become out of date, but were correct at the time of writing.

www.redkites.net – Includes useful information about reading wing-tags and the meaning of the different tag colours used, as well as much else Red Kite related.

Natural England (previously English Nature) – A research report about the reintroduction project is available as a free download: http://publications. naturalengland.org.uk/publication/127035.

Campaign for Responsible Rodenticide Use (CRRU) – Invaluable information on the threat posed to wildlife and advice on the safe use of rat poisons: www. thinkwildlife.org/code-of-best-practice/crru-code.

Scottish Natural Heritage – Information about the reintroduction programme north of the border, including a report on the impacts of wildlife crime and wind turbines: www.nature.scot/snh-commissioned-report-904-population -modelling-north-scotland-red-kites-relation-cumulative.

RSPB – General information about the Red Kite and its conservation: www.rspb. org.uk/birds-and-wildlife/wildlife-guides/bird-a-z/red-kite.

Chilterns AONB – Information and fact sheets about the Red Kite, with a focus on the local population in the Chiltern Hills: https://chilternsaonb.org/about -chilterns/red-kites.html.

Yorkshire Red Kites – The best place for information about the expanding Yorkshire Red Kite population: http://yorkshireredkites.net.

Welsh Kite Trust – Information on Red Kites and other raptors in Wales, including detailed newsletters available to download: https://welshkitetrust.wales.

Golden Eagle Trust – As the name suggests, this group is not involved solely with kites, but it has taken a lead role in the Red Kite reintroduction work in the Republic of Ireland: http://goldeneagle.ie.

Finally, online video-sharing sites such as YouTube are not to be underestimated as a source of entertainment, and even enlightenment, in relation to Red Kite behaviour. There are hundreds of videos involving Red Kites, ranging from huge gatherings at feeding stations (Gigrin Farm, Bellymack Hill Farm and others) to feeding in village gardens, activity at the nest, skirmishes with other raptors and corvids, and Red Kites flying with gliders and light aircraft.

SPECIES MENTIONED IN THE TEXT

BIRDS

American Black Vulture *Coragyps atratus*
Black-headed Gull *Chroicocephalus ridibundus*
Black Kite *Milvus migrans*
(Common) Buzzard *Buteo buteo*
Carrion Crow *Corvus corone*
Chicken *Gallus gallus domesticus*
Eagle Owl *Bubo bubo*
Fulmar *Fulmarus glacialis*
Golden Eagle *Aquila chrysaetos*
Griffon Vulture *Gyps fulvus*
Hen Harrier *Circus cyaneus*
(Grey) Heron *Ardea cinerea*
Jackdaw *Coloeus monedula*
Jay *Garrulus glandarius*
Kestrel *Falco tinnunculus*
Long-tailed Tit *Aegithalos caudatus*
Magpie *Pica pica*
Montagu's Harrier *Circus pygargus*
Mute Swan *Cygnus olor*
Peregrine Falcon *Falco peregrinus*
Pheasant *Phasianus colchicus*
Raven *Corvus corax*
Red Grouse *Lagopus lagopus*
Red Kite *Milvus milvus*
Red-legged Partridge *Alectoris rufa*
Rook *Corvus frugilegus*
Skylark *Alauda arvensis*
Tawny Owl *Strix aluco*
White-tailed Eagle *Haliaeetus albicilla*
Woodpigeon *Columba palumbus*
Wren *Troglodytes troglodytes*

OTHER VERTEBRATES

Badger *Meles meles*
Brown Rat *Rattus norvegicus*
(Domestic) Cat *Felis catus*
(Red) Fox *Vulpes vulpes*
Grey Squirrel *Sciurus carolinensis*
(Brown) Hare *Lepus europaeus*
Pine Marten *Martes martes*
Polecat *Mustela putorius*
Rabbit *Oryctolagus cuniculus*
Red Squirrel *Sciurus vulgaris*
(Atlantic) Salmon *Salmo salar*
Sheep *Ovis aries*
Stoat *Mustela erminea*
Weasel *Mustela nivalis*

INVERTEBRATES

Brown Argus *Aricia agestis*
Chalkhill Blue *Lysandra coridon*
Earthworm *Lumbricus terrestris* (and similar species)
Silver-spotted Skipper *Hesperia comma*

PLANTS

Autumn Gentian *Gentianella amarella*
Clustered Bellflower *Campanula glomerata*
Eyebright *Euphrasia* sp.
Field Scabious *Knautia arvensis*
Harebell *Campanula rotundifolia*
Marjoram *Origanum vulgare*
Round-headed Rampion *Phyteuma orbiculare*

INDEX